INTERNATIONAL ANIMAL RESEARCH REGULATIONS

Impact on Neuroscience Research

WORKSHOP SUMMARY

Diana E. Pankevich, Theresa M. Wizemann, Anne-Marie Mazza, and
Bruce M. Altevogt, *Rapporteurs*

Forum on Neuroscience and Nervous System Disorders
Board on Health Sciences Policy

Committee on Science, Technology, and Law
Policy and Global Affairs Division

Institute for Laboratory Animal Research
Division on Earth and Life Sciences

INSTITUTE OF MEDICINE *AND*
NATIONAL RESEARCH COUNCIL
OF THE NATIONAL ACADEMIES

THE NATIONAL ACADEMIES PRESS
Washington, D.C.
www.nap.edu

THE NATIONAL ACADEMIES PRESS • 500 Fifth Street, NW • Washington, DC 20001

This project was supported by contracts between the National Academy of Sciences and the Alzheimer's Association; AstraZeneca Pharmaceuticals, Inc.; CeNeRx Biopharma; the Department of Health and Human Services' National Institutes of Health (NIH, Contract Nos. N01-OD-4-2139) through the National Institute on Aging, National Institute on Alcohol Abuse and Alcoholism, National Institute on Drug Abuse, National Eye Institute, NIH Blueprint for Neuroscience Research, National Institute of Mental Health, and National Institute of Neurological Disorders and Stroke; Eli Lilly and Company; Foundation for the National Institutes of Health; GE Healthcare, Inc.; GlaxoSmithKline, Inc.; Johnson & Johnson Pharmaceutical Research and Development, LLC; Lundbeck Research USA; Merck Research Laboratories; The Michael J. Fox Foundation for Parkinson's Research; the National Multiple Sclerosis Society; the National Science Foundation (Contract No. OIA-0753701); Pfizer Inc.; and the Society for Neuroscience. The views presented in this publication are those of the editors and attributing authors and do not necessarily reflect the view of the organizations or agencies that provided support for this project.

International Standard Book Number-13: 978-0-309-25208-9
International Standard Book Number-10: 0-309-25208-3

Additional copies of this report are available from the National Academies Press, 500 Fifth Street, NW, Keck 360, Washington, DC 20001; (800) 624-6242 or (202) 334-3313; http://www.nap.edu.

For more information about the Institute of Medicine, visit the IOM home page at: **www.iom.edu.**

The serpent has been a symbol of long life, healing, and knowledge among almost all cultures and religions since the beginning of recorded history. The serpent adopted as a logotype by the Institute of Medicine is a relief carving from ancient Greece, now held by the Staatliche Museen in Berlin.

Suggested citation: IOM (Institute of Medicine) and NRC (National Research Council). 2012. *International animal research regulations: Impact on neuroscience research: Workshop summary.* Washington, DC: The National Academies Press.

THE NATIONAL ACADEMIES
Advisers to the Nation on Science, Engineering, and Medicine

The **National Academy of Sciences** is a private, nonprofit, self-perpetuating society of distinguished scholars engaged in scientific and engineering research, dedicated to the furtherance of science and technology and to their use for the general welfare. Upon the authority of the charter granted to it by the Congress in 1863, the Academy has a mandate that requires it to advise the federal government on scientific and technical matters. Dr. Ralph J. Cicerone is president of the National Academy of Sciences.

The **National Academy of Engineering** was established in 1964, under the charter of the National Academy of Sciences, as a parallel organization of outstanding engineers. It is autonomous in its administration and in the selection of its members, sharing with the National Academy of Sciences the responsibility for advising the federal government. The National Academy of Engineering also sponsors engineering programs aimed at meeting national needs, encourages education and research, and recognizes the superior achievements of engineers. Dr. Charles M. Vest is president of the National Academy of Engineering.

The **Institute of Medicine** was established in 1970 by the National Academy of Sciences to secure the services of eminent members of appropriate professions in the examination of policy matters pertaining to the health of the public. The Institute acts under the responsibility given to the National Academy of Sciences by its congressional charter to be an adviser to the federal government and, upon its own initiative, to identify issues of medical care, research, and education. Dr. Harvey V. Fineberg is president of the Institute of Medicine.

The **National Research Council** was organized by the National Academy of Sciences in 1916 to associate the broad community of science and technology with the Academy's purposes of furthering knowledge and advising the federal government. Functioning in accordance with general policies determined by the Academy, the Council has become the principal operating agency of both the National Academy of Sciences and the National Academy of Engineering in providing services to the government, the public, and the scientific and engineering communities. The Council is administered jointly by both Academies and the Institute of Medicine. Dr. Ralph J. Cicerone and Dr. Charles M. Vest are chair and vice chair, respectively, of the National Research Council.

www.national-academies.org

U.S. AND EUROPEAN ANIMAL RESEARCH REGULATIONS: IMPACT ON NEUROSCIENCE RESEARCH PLANNING COMMITTEE[*]

COLIN BLAKEMORE (*Co-Chair*), Oxford University
ARTHUR SUSSMAN (*Co-Chair*), MacArthur Foundation
ROBERTO CAMINITI, University of Rome
JUDY MacARTHUR CLARK, Animals Scientific Procedures Inspectorate
RICHARD CUPP, Pepperdine Law School
MARGARET LANDI, GlaxoSmithKline
ALAN LESHNER, American Association for the Advancement of Science
RICHARD NAKAMURA, National Institute of Mental Health
TIMO NEVALAINEN, University of Eastern Finland
MICHAEL OBERDORFER, National Eye Institute (Retired)
FRANKIE TRULL, Foundation for Biomedical Research

Study Staff

BRUCE M. ALTEVOGT, Project Director, Institute of Medicine
DIANA E. PANKEVICH, Associate Program Officer, Institute of Medicine
LEILA AFSHAR, Research Associate, Institute of Medicine (until August 2011)
LORA K. TAYLOR, Senior Project Assistant, Institute of Medicine (until December 2011)
ANNE-MARIE MAZZA, Director, Committee on Science Technology and Law, National Research Council
LIDA ANESTHEDIOU, Senior Program Officer, Institute of Laboratory Animal Research, National Research Council

---

[*] Institute of Medicine planning committees are solely responsible for organizing the workshop, identifying topics, and choosing speakers. The responsibility for the published workshop summary rests with the workshop rapporteurs and the institution.

FORUM ON NEUROSCIENCE AND
NERVOUS SYSTEM DISORDERS*

---

* Institute of Medicine forums and roundtables do not issue, review, or approve individual documents. The responsibility for the published workshop summary rests with the workshop rapporteurs and the institution.

Reviewers

This report has been reviewed in draft form by individuals chosen for their diverse perspectives and technical expertise, in accordance with procedures approved by the National Research Council's Report Review Committee. The purpose of this independent review is to provide candid and critical comments that will assist the institution in making its published report as sound as possible and to ensure that the report meets institutional standards for objectivity, evidence, and responsiveness to the study charge. The review comments and draft manuscript remain confidential to protect the integrity of the process. We wish to thank the following individuals for their review of this report:

Floyd Bloom, The Scripps Research Institute
Barbara Davies, Understanding Animal Research
Sharon Juliano, Uniformed Services University of the Health Sciences
Emily McIvor, Humane Society International
Robert Wurtz, National Eye Institute
Stuart Zola, Yerkes Regional Primate Research Center

Although the reviewers listed above have provided many constructive comments and suggestions, they did not see the final draft of the report before its release. The review of this report was overseen by **Caswell Evans**, associate dean of the College of Dentistry, University of Illinois at Chicago, and **Joseph T. Coyle**, Eben S. Draper Professor of Psychiatry and of Neuroscience, Harvard Medical School. Appointed by the Institute of

Medicine, they were responsible for making certain that an independent examination of this report was carried out in accordance with institutional procedures and that all review comments were carefully considered. Responsibility for the final content of this report rests entirely with the authoring committee and the institution.

Contents

1

Introduction and Overview[1]

Animals are widely used in neuroscience research to explore normal and abnormal biological mechanisms of nervous system function, to identify the genetic basis of disease states, and to provide models of human disorders and diseases for the development of new treatments. Numerous laws, policies, and regulations are in place governing the use of animals in research. These measures are intended to ensure the humane care and use of animals, including the implementation of practical steps to use the smallest number of animals necessary to achieve significant results while minimizing pain and distress. Many animal care and use issues are generic to all types of biomedical research; however, animal regulations have implications specific to neuroscience research.

To consider these issues from a global perspective, the Institute of Medicine Forum on Neuroscience and Nervous System Disorders, in collaboration with the National Research Council Committee on Science, Technology, and Law and the Institute for Laboratory Animal Research,

[1] This workshop was organized by an independent planning committee whose role was limited to identification of topics and speakers. This workshop summary was prepared by the rapporteurs as a factual summary of the presentations and discussions that took place at the workshop. Statements, recommendations, and opinions expressed are those of individual presenters and participants, and are not necessarily endorsed or verified by the Forums or the National Academies, and they should not be construed as reflecting any group consensus. Furthermore, although the current affiliations of speakers and participants are noted in the report, many qualified their comments as being based on personal experience over the course of a career and are not being presented formally on behalf of their organizations (unless specifically noted).

convened the workshop "U.S. and European Animal Research Regulations: Impact on Neuroscience Research." Held at the Kavli Royal Society International Centre in Buckinghamshire, UK, on July 26-27, 2011, the workshop brought together neuroscientists, legal scholars, administrators, and other key stakeholders to discuss current and emerging trends in animal regulations as they apply to the neurosciences. As outlined by co-chairs Colin Blakemore, professor of neuroscience at the University of Oxford, and Arthur Sussman, of the University of Chicago Law School, the workshop was designed to

- identify and discuss current international animal use regulations;
- examine the implications of current policies on the research enterprise, including the impact of disparate policies;
- discuss developments in law school curriculums, animal law practice, and activity in the courts that may affect the use of animals in research;
- explore the reasons for the establishment of specific regulations; and
- discuss opportunities for harmonization of regulations and/or the development of global core principles.

ANIMAL RESEARCH IN THE NEUROSCIENCES: INTRODUCTION BY COLIN BLAKEMORE

As background for the workshop discussions, Blakemore highlighted some of the current issues surrounding the use of animals in neuroscience research. In research involving animals, he acknowledged a necessary tension between the desire to benefit from the advances in knowledge that accrue from studies in animals and the desire to avoid deliberate harm to the animals. Opinions on the use of animals in research are polarized. Researchers, clinicians, and institutions that support animal research, along with a portion of the general public, accept its importance for progress in medicine. The principal argument for using animals in biomedical research states that it is ethically more acceptable than neglecting the suffering of the sick (human or animal, as animal research also benefits animals). Some individuals and organizations, however, oppose animal research on ethical grounds; they contend that humans should not benefit from animal suffering (a deontological argument: actions are either intrinsically right or wrong, regardless of the consequences). Some challenge the validity of animal models and the unreliability of treatments developed through the study of non-human species (a utilitarian argument: the correct action is the one that maximizes the overall good, specifically considering the consequences). Others claim that alternative methods to animal use are already available or could be available with increased efforts to develop them.

BOX 1-1
Particular Issues Surrounding Animal Use in Neuroscience

- Models of nervous system disease may include behavioral and cognitive phenotypes that have the potential to result in suffering:
  - ○ Research on conditions such as addiction, depression, anxiety, and fear may be problematic.
- Pain, which is normally avoided in the design of experiments, is an important topic of study.
- Modification of sensory experience may be considered a form of suffering.
- Non-human primate use raises concerns due to costs and public perception.
- Research may involve invasive methodology, restriction or control of food or water intake, and/or prolonged or repetitive procedures

SOURCE: Blakemore presentation.

Animal Use Issues Specific to Neuroscience

Blakemore highlighted several issues associated with the use of animals that are specific to neuroscience research, such as the use of non-human primates, pain as a topic of study, and the use of invasive methodologies (Box 1-1). Non-human primates, due to their close phylogenetic relatedness to humans, make them the preferred species to study issues such as fine motor control, high-level cognitive functions, and decision making. This close evolutionary proximity to humans increases scrutiny of the use of non-human primates and raises special concerns, including public attitude about their use, supply issues, and costs. The 2006 Weatherall Report[2] concluded that there is scientific justification for the carefully regulated use of non-human primates when there is no other way to address clearly defined questions, including those raised by certain neuroscience studies (MRC, 2006).

Another issue is the use of genetically modified animals. Modification of genes that regulate the nervous system and neurologic development can produce a particular phenotype with behavioral and cognitive consequences. The impact of the phenotype itself, in terms of suffering, must be taken into account even before considering the impact of procedures to be carried out on genetically modified animals. The introduction of human genetic mate-

---

[2] Note that a comprehensive 5-year follow-up review of the quality and impact of primate research has been published. *Review of Research Using Non-Human Primates: Report of a Panel Chaired by Professor Sir Patrick Bateson FRS* is available at http://www.mrc.ac.uk/Utilities/Documentrecord/index.htm?d=MRC008083.

rial into animals is another topic of much discussion. For example, a variety of mouse models for Huntington's disease, a neurodegenerative disorder, incorporate a portion of the human Huntington gene. Blakemore referred participants to an Academy of Medical Sciences (UK) report on the use of animals containing human genetic material that was released the same week as the workshop (AMS, 2011).

Neuroimaging is increasingly being used in animal studies. While imaging is noninvasive, such studies are generally longitudinal, involving repetitive procedures. It is not always clear, Blakemore noted, that imaging is preferable to invasive methodologies.

A Framework for Research on Animals

Blakemore suggested there is a strong need for a regulatory framework that is ethically secure, consistent, legally strong, and defensible but not so overly burdensome that it impedes scientific progress. Such a framework might include strict requirements for and a commitment to high-quality welfare and good husbandry; good recordkeeping; transparency and accountability; provisions for public engagement; certification of researchers so that their skills are documented and controlled; and a system for approval of individual projects based on cost-benefit analysis.

Cost-benefit analysis, while theoretically straightforward, is challenging to apply to animal research. By definition, research involves the unknown, and potential future benefits cannot definitively be known in advance. In contrast, the immediate costs relative to the suffering of animals can be determined before and during animal experiments. Therefore, individuals continually weigh *potential* benefits against *definite* costs, Blakemore said.

A 1999 poll published in the New Scientist found that 83 percent of those surveyed would support research on mice to study childhood leukemia if there were no pain involved (Aldhous et al., 1999). If pain or death was involved, 63 and 69 percent respectively still supported the research. In a 2000 poll by the UK-based Ipsos Market and Opinion Research International (Ipsos MORI) firm, only 32 percent of those surveyed supported research on animals in general if there were no alternatives, but when the same group was asked whether they would accept the use of animals for medical research if there was no unnecessary suffering, 84 percent agreed (Ipsos MORI, 2000). This, Blakemore said, shows that people are performing quite complex personal calculations and shifting their views depending on the perceived costs and benefits.

Opinion has changed dramatically and progressively over the past 10 years in the United Kingdom, Blakemore noted. The greatest change in opinion was between 1999 and 2004. Since 2004, polls conducted by Ipsos MORI have consistently shown that 87 percent of the public conditionally accepts the use of animals in research for medical benefit, if suffering is minimized and/or there is no alternative to the use of animals (Ipsos MORI, 2010).

A key component of the shifting views toward animal research in the United Kingdom has been due to increased openness and public engagement. It is important, Blakemore stressed, that scientists themselves speak out to win the trust of the public, politicians, and the media. Ipsos MORI polls show that the majority of the public trusts scientists to tell them the truth, yet scientists do not normally engage with the public to provide information about their research (Ipsos MORI, 2008). Blakemore noted that in recent years, against the backdrop of political support, including new legislation to prevent violence, there has been an increase in the willingness of researchers to talk openly about their work. In addition, there was growing public support of animal research with groups such as Pro-Test[3] holding rallies in support of animal testing in medical research. In reports of advances resulting from research, more institutions are openly identifying the animal species used in the research, which, Blakemore noted, has had a positive impact on public opinion.

ORGANIZATION OF THE WORKSHOP AND REPORT

Following the overview of issues presented by Blakemore, the workshop considered current and emerging international regulations governing animal research and the impact of legal trends, including animal rights laws, Freedom of Information requests, and state "sunshine laws," on the use of animals in research (Chapters 2 and 3). The next session of the workshop focused on the implications of these laws, regulations, and policies for neuroscience research, and considered case studies applying the "3Rs" (replacement, refinement, and reduction) to neuroscience research (Chapters 4 and 5). The final portion of the workshop focused on engaging the public, politicians, and the media in animal research issues, and developing core principles for regulating the use of animals in research (Chapters 6 and 7). In the closing session, session chairs identified what they viewed as the key points that emerged (Chapter 8 and summarized below).

Highlights of Workshop Sessions

Session chairs noted key points that emerged during workshop presentations and discussions:

---

[3] In February 2011, 5 years after it formed, Pro-Test wound up its UK operations because it had successfully met its goals. See http://www.pro-test.org.uk/.

- **Regulatory Harmonization (Session I):** Animal research regulations in the United States and the European Union are more similar than different. International collaborations are helping to influence new regulations, raise standards in emerging regions (e.g., Asia, South America), and contribute to global harmonization.
- **Administrative Burden (Sessions I and III):** Regulatory systems have a variety of costs, including financial costs, the costs of increased oversight for regulators, and the costs of lost research time for scientists. Appropriate measures of the success of animal welfare regulations can be useful because it is unclear whether increased costs and burdens result in improved animal welfare.
- **Legal Trends (Session II):** The effect of increased attention on animal rights laws is unclear. Freedom of Information requests and state sunshine laws are used in the United States to allow the public to access detailed information about the use of animals in research. The effect of these laws on animal research is not yet known.
- **Non-Human Primates in Neuroscience (Session III):** Non-human primates continue to be used in biomedical research, including neuroscience research. Such studies complement in vitro studies, in silico modeling, human brain imaging, and parallel investigations in rodents and other species.
- **Data Sharing (Session IV):** Systematic reviews of preclinical data could potentially support the 3Rs (replacement, refinement, and reduction), improve the quality and value of animal studies, and better inform clinical trials. Research might benefit if preclinical animal data are more accessible, including negative data, primary data, and precompetitive data.
- **Engaging the Public (Session V):** Communication between the scientific community and the public, the media, and policy makers about the role and welfare of animals in neuroscience research is critical. In some countries, public engagement and education can impact the public view of the use of animals in research.
- **Aligning Core Principles to Achieve Consistent Animal Care and Use Outcomes (Sessions I and VI):** Animal research regulations might benefit from a careful balance of *quality science, animal welfare,* and *public confidence.* Animal welfare can be considered together with scientific goals and the larger needs of society. Alignment of animal research principles can be achieved independent of differing policies or practices. Core principles governing how animal studies might be conducted are the same for any discipline, including neuroscience.

2

The Evolving Regulatory Environment

Judy MacArthur Clark, chief inspector of the Animals Scientific Procedures Inspectorate at the UK Home Office and session chair, described regulatory balance as the overlap of *scientific quality, animal welfare,* and *public confidence.* MacArthur Clark observed that regulations should promote high-quality science and ensure that animal suffering is minimized without developing bureaucratic systems that are obstructive. Extensive evidence demonstrates that the quality of science is impacted by the welfare of the animals. The public wants to benefit from scientific advances, but also wants to be reassured research does not impose unnecessary suffering on animals, said MacArthur Clark. The nature of this balance can vary among countries due to differences in culture, economy, religion, and social factors.

To explore similarities and differences across countries and regions, four invited speakers from Europe, North America, Asia, and South America described current regulations in their regions and emerging issues surrounding animal research.

EUROPE: EUROPEAN UNION

Karin Blumer, of Novartis, Switzerland, described pan-European legislation, that is, legislation that applies to all 27 member states of the European Union (EU), but not necessarily to all countries of the European continent.

History of European Animal Welfare Legislation

Historically, animal research regulations across Europe have been fragmented. In the early 1800s, the first animal welfare legislation was introduced in Great Britain, followed by Saxony and Germany; however, these laws were not specific to laboratory use of animals. The first law specifically governing the use of animals in laboratories was not enacted until the late 1800s in Great Britain.

In the twentieth century, significant differences remained in animal welfare awareness and legal protection across Europe. Many countries enacted laboratory animal welfare legislation during the mid- to late 1980s. In 1986, Europe, as a political union (then the European Economic Community), approved Directive 86/609, which very specifically governed the use of laboratory animals (European Communities and Office for Official Publications, 1986). Blumer noted that this legislation set minimum standards across member states while allowing for stricter national-level legislation. However, in 2000, national animal welfare legislation still varied widely across the European Union.

Given the significant advances in biomedical technology, the addition of new member states whose animal welfare legislation was rudimentary or nonexistent, and increased public sensitivities, stakeholders concluded that revisions to EU Directive 86/609 were needed. One of the criticisms of EU Directive 86/609 was that it did not include strong guidance on housing and care, or provisions for genetic modifications of animals. Revision of the law began in 2002, and in 2010, EU Directive 2010/63 was adopted, updating and replacing Directive 86/609 (European Union, 2010). The deadline for adoption and transposition of the directive is January 2013.

EU Directive 2010/63

The new directive is a complex document, Blumer said, structured as an introduction with 56 "recitals" explaining the rationale and objectives of the law, followed by 6 chapters defining provisions, procedures, and authorizations, and 8 annexes providing additional detail (Box 2-1).

The main areas of focus are

- harmonization among EU member states;
- expansion of the legislative scope (e.g., more species, earlier stages);
- a push for the implementation of the 3Rs (replacement, refinement, and reduction);
- authorization of projects;
- limits on individual animal exposure (including upper pain limits and limitations on the reuse of animals);

BOX 2-1
Key Features of European Union Directive 2010/63

Chapter I (General Provisions)
- Widened scope: Specific invertebrates and fetuses in last trimester of development; animals in basic research, education, and training (Art. 2).
- Formal introduction of 3Rs (reduction, refinement, and replacement) as guiding principles (Art. 4).
- Limitation of acceptable methods of sacrifice (Art. 6, Annex IV).

Chapter II (Provisions for Certain Animals)
- Restricted use of endangered species (Art. 7).
- Restricted use of non-human primates, ban on use of Great Apes (Art. 8).
- Purpose-bred requirement for most commonly used lab species; F2 requirement for non-human primates (Art. 10, Annex I).

Chapter III (Procedures)
- Mandatory use of alternatives, reduction, and refinement (Art. 13).
- Severity classification system, ban on severe studies (Art. 15).
- Reuse limitations (Art. 16).

Chapter IV (Authorization)
- Competence of personnel, institutional animal welfare person, designated veterinarian, and animal welfare body (Art. 23, 24, 25, 27).
- Tasks of animal welfare body (Art. 27).
- Breeding strategy for non-human primates (Art. 28).
- Care and accommodation (Art. 33).
- Inspections, controls of member states inspections (Art. 34, 35).
- Project authorization, application, evaluation (including ethical considerations), retrospective assessment, granting of authorization, simplified administration procedure (Art. 36).
- Non-technical project summaries (Art. 43).

Chapter V (Avoidance of Duplication and Alternatives)
- Mutual data acceptance of member states (Art. 46).
- Union Reference Laboratory (Art. 48).
- National committees for laboratory animal protection (Art. 49).

Chapter VI (Final Provisions)
- Reporting obligations for member states (Art. 54).
- Safeguard clauses (Art. 55)—Great Apes, most severe studies.
- Commission report to the European Parliament and the Council—every 5 years after 2019 (Art. 57).
- Five-year reviews (2017) with special focus on advancements of alternatives—specifically for non-human primates (Art. 58).

SOURCE: Blumer presentation.

- highly specific regulations for certain species including Great Apes and non-human primates;
- increased transparency within institutions and to the public; and
- continuous review of laws and regulations.

One of the biggest advancements for animal welfare, Blumer said, is that fundamental principles of care and accommodation will now be harmonized on a pan-European level, which means that all member states will be required to use the same housing and care standards.

Emerging Trends

Blumer noted that an emerging trend in the European Union over the past 10 years is the increased involvement of the lay public in issues regarding animal research regulation. The societal call for special status or "rights" for select animals such as Great Apes or companion animals is increasingly reflected in the legislative process. Blumer also noted a growing recognition that policy makers no longer recognize or understand the essential nature of science and research because science is so complex and great "breakthroughs" have been so limited. A related trend is that many policy makers fail to appreciate the essential importance of basic science to applied research and innovation. Finally, Blumer pointed out an emerging reductionist approach to the 3Rs, with the primary focus often only on replacement.

NORTH AMERICA: UNITED STATES

Taylor Bennett, senior scientific advisor for the National Association for Biomedical Research (NABR), referred participants to the *Guide for the Care and Use of Laboratory Animals* which summarizes the U.S. animal regulatory environment. The guide says, "The use of laboratory animals is governed by an interrelated, dynamic system of regulations, policies, guidelines, and procedures" (NRC, 2010). This oversight system is composed of both activities mandated by law or required as a condition of funding and activities that an individual or institution voluntarily adhere to as part of their overall commitment to research and academic excellence (Box 2-2).

Animal Welfare Act

The Animal Welfare Act[1] passed in 1966 and has been amended six times, most recently in 2008. The Act empowers the U.S. Depart-

---

[1] See http://awic.nal.usda.gov/nal_display/index.php?info_center=3&tax_level=3&tax_subject=182&topic_id=1118&level3_id=6735&level4_id=0&level5_id=0&placement_default=0.

BOX 2-2
Components of the U.S. Animal Research Oversight System

Mandatory
- Animal Welfare Act, enforced by U.S. Department of Agriculture (USDA).
- Public Health Service (PHS) Policy on Humane Care and Use of Laboratory Animals, administered by the Office of Laboratory Animal Welfare (OLAW) of the National Institutes of Health (NIH).
- Good Laboratory Practices (GLPs) regulations of the U.S. Food and Drug Administration (FDA).
- Requirements set by private funding agencies.

Voluntary
- Accreditation:
  - Association for the Assessment and Accreditation of Laboratory Animal Care International (AAALAC).
- Standards set and maintained by individual users.

SOURCE: Bennett presentation.

---

ment of Agriculture (USDA) to develop definitions, regulations, and standards for the care and use of animals, including laboratory animals. The USDA licenses animal dealers, registers research institutions (~1,100 in the United States), requires certain recordkeeping, and enforces the law through unannounced inspections. The Act defines "animal" as a warm blooded mammal (excluding birds, mice, and rats) raised for research and farm animals used for agricultural purposes. Animal welfare regulations can be modified through amendments to the Act itself. In addition, the Secretary of Agriculture may propose additions or changes to the existing regulations by publication of a Proposed Rule in the *Federal Register*.

Bennett explained that the Act prohibits the promulgation of rules, regulations, or orders that would interfere with the conduct of actual research. Determination of what constitutes actual research is left to the discretion of the research facility. Rules, regulations, or orders may be added as they relate to areas covered by the program of adequate veterinary care and areas that ensure that professionally acceptable standards governing the care, treatment, and use of animals are followed by the research facility during actual research or experimentation.

Animal welfare regulations establish certain institutional responsibilities, including the appointment of an onsite Institutional Animal Care and Use Committee (IACUC) and a program of adequate veterinary care. The regulations also cover the training of qualified personnel, recordkeeping, and annual reports.

Institutional Animal Care and Use Committee

A key function of the IACUC is the semi-annual inspection and review of animal facilities, investigators' laboratories, and overall management practices (the "Animal Care and Use Program"). Reports of these investigations are provided to responsible institutional officials who oversee the program and, if necessary, include a specified time frame for correcting any deficiencies. Failure to correct deficiencies are noted and reported to the appropriate regulatory agency. The IACUC also reviews concerns raised by both internal and external groups and has the authority to make recommendations to institutional officials on any aspect of the Animal Care and Use Program.

The majority of the IACUC time is dedicated to reviewing and approving or requiring modification of research protocols involving the use of animals. Part of that review is to ensure that personnel are properly trained and that the investigator adheres to the principles of the 3Rs in terms of justification of alternative methodology and assurance of unnecessary duplication. The IACUC has the authority to suspend research activity and must report any such suspensions to regulatory and funding agencies.

Attending Veterinarian and Adequate Veterinary Care Program

Institutions conducting research involving animals are required to employ an attending veterinarian with the authority to ensure the provision of adequate veterinary care and to oversee the adequacy of other aspects of the animal care and use program. The veterinarian also is a voting member of the IACUC.

Each research facility is required to provide the attending veterinarian with the necessary resources to manage an effective program of veterinary care, including facilities, personnel, equipment, and services. The veterinarian needs to be able to implement appropriate methods and systems to prevent, control, diagnose, and treat diseases and injuries through daily observation of all the animals as well as through emergency care.

Finally, the veterinarian must be provided with the necessary resources to be able to provide guidance to investigators and other personnel regarding handling, immobilization, anesthesia, analgesia, tranquilization, and euthanasia, and ensuring adequate pre- and postprocedural care in accordance with current established veterinary medical and nursing procedures.

Public Health Service (PHS) Policy

The PHS Policy on Humane Care and Use of Laboratory Animals[2] is intended to implement and supplement the U.S. Government Principles for the Utilization and Care of Vertebrate Animals Used in Testing, Research, and Training.[3] The PHS requires the institutions it funds to follow the *Guide for the Care and Use of Laboratory Animals.*

While the USDA regulations exclude rats, mice, and birds raised for research, the PHS Policy covers all vertebrate animals. The PHS process is self-regulatory, the IACUC composition is slightly different, and the PHS has an Animal Welfare Assurance process, whereas the USDA has a registration process. Another difference between the two agencies is that the USDA conducts unannounced inspections while PHS only does inspections for cause. Together, however, the USDA regulations and standards and the PHS policy provide broad oversight of key animal care and use issues.

Emerging Trends

Bennett highlighted four emerging trends in U.S. regulation of the use of animals in research. First, institutional "downstreaming" due to decreasing budgets at many research institutions is increasing the administrative burden of animal regulation. Bennett indicated that there is anecdotal evidence that departments and individual investigators are spending more time on administrative issues associated with the use of animals. Another trend is the USDA shift away from an education focus to an enforcement focus, leading to increased citations, fines, and animal-use stipulations. The increasing use of Freedom of Information requests is leading to a growing administrative burden to fulfill requests and increasing visibility of individual investigators as identities are disclosed.[4] Finally, as financial resources decline, the cost of assuring regulatory compliance is being passed on to the investigators in the form of increased per diems and service charges, reducing the resources available for research itself, Bennett asserted.

ASIA: CHINA

Jianfei Wang, director of laboratory animal science at GlaxoSmithKline Research and Development Center, China, provided a high-level summary of laboratory animal regulation in Asia (Box 2-3), followed by specific ex-

---

[2] See http://grants.nih.gov/grants/olaw/references/phspol.htm.
[3] The U.S. Government Principles for the Utilization and Care of Vertebrate Animals Used in Testing, Research, and Training are discussed further by Brown in Chapter 7 of this document.
[4] Discussed further in Chapter 3.

BOX 2-3
Laboratory Animal Welfare Regulations in Asia

Japan
- Guideline for Proper Conduct of Animal Experiments passed in 2000, revised in 2006.
- Institutional Animal Care and Use Committee (IACUC) responsibility for oversight and incorporation of 3Rs (reduction, refinement, and replacement).
- See http://www.scj.go.jp/ja/info/kohyo/pdf/kohyo-20-k16-2e.pdf.

Korea
- Animal protection law passed in 2007.
- Ethical review committee along with consideration of alternative models.
- See http://www.koreananimals.org/animals/apl/2007apl.htm.

Singapore
- An extensive set of guidelines released in 2004.
- Based on the principles of the 3Rs, IACUC responsibility for oversight.
- See http://www.ava.gov.sg/AnimalsPetSector/CareAndUseAnimalsForScientific Purp/#naclar.

India
- Prevention of Cruelty to Animals (PCA) issued in 1960.
- Committee for the Purpose of Control and Supervision of Experiments on Animals (CPCSEA), Institutional Animal Ethics Committee (IAEC).
- See http://www.aaalac.org/resources/CPCSEA_Conference_IAEC_SOP.pdf.

China
- Guideline on Administration of Laboratory Animals released in 1988.
- Humane treatment of laboratory animals and IACUC review.
- See http://www.lascn.net/policy/Index.html.

SOURCE: Wang presentation.

---

amples from China. Asia is very diverse, he noted, and there is no pan-Asian union comparable to the Euroepan Union.

Laboratory Animal Science and Regulations in China

Thirty years ago, Wang said, there was essentially no concept of laboratory animal science in China. However, this has changed in recent years. Under the direction of the Ministry of Science and Technology (MOST), the Provincial Department of Science and Technology (PDST) handles organizational licensing and inspections through Provincial Administrative

Offices. The National Monitoring Center oversees the quality of animal programs and facilities through Provincial Monitoring Units. Additionally, the Chinese Association for Laboratory Animal Science (CALAS) promotes laboratory animal science education and training.

Wang noted that the Chinese government has issued more than 100 regulatory standards for topics such as microorganism control, environment and housing facilities, genetic quality control, and diet and nutrition. The purpose of these national standards is to ensure both the quality of laboratory animals and the scientific knowledge derived from these animals. To promote laboratory science in China, the government established eight laboratory animal resource centers. There also are graduate and undergraduate programs in laboratory animal science at universities and medical, veterinary, pharmacy, and biotechnology schools.

Wang mentioned three legal milestones governing laboratory animal science in China: the 1988 Statute of Laboratory Animal Administration, the 2001 Regulation on the Management of Laboratory Animal License System, and the 2006 Guideline on Humane Treatment of Laboratory Animals.

Guideline on Humane Treatment of Laboratory Animals

The Guideline on Humane Treatment of Laboratory Animals is the country's first broad animal welfare regulation and is aligned with practices in countries/regions such as the United States and the European Union. The notion of animal welfare is well accepted by Chinese scientists and the government, Wang noted. The guideline covers animal welfare from procurement through completion of a research project. IACUC review is required and the principles of 3Rs must be incorporated into the experimental design. A proper animal environment, husbandry, and care must be provided, as well as adequate veterinary care. Pain and distress are a particular focus of the guidelines, so humane endpoints must be established and animals must be properly euthanized.

Factors Influencing Quality of Animal Care and Use in China

Scientists in China have come to realize that good animal welfare is necessary for good science, Wang observed. Regulatory practice has been influenced by an increasing number of Chinese scientists trained overseas who have returned to China with an understanding of the practices and protocols of other countries. Additional factors causing change in China's regulatory practices include increased international academic collaboration and participation in international conferences where scientists share animal welfare information.

The Chinese culture of embracing traditional virtues such as compassion for living things, Wang explained, also influences animal care because cruelty to animals creates a very negative public image. Another positive influence on quality of animal care and use in China is the presence of multinational pharmaceutical and contract research organizations. While there are diverse animal study requirements to satisfy the needs of drug discovery, development, and safety assessment, animal welfare needs appear to be the same. The number of institutions in the Pacific Rim that are accredited by the Association for the Assessment and Accreditation of Laboratory Animal Care International (AAALAC) is increasing, including several institutions in China.

Economic concerns also influence animal care and use. China is one of the largest agricultural countries in the world, yet it faces severe restraints on trade in animal products due to concerns over quality standards and animal protection. Wang noted that the Chinese government is beginning to realize that animal welfare is linked to economic development. Animal resources in China, however, are not as abundant as in the United States and veterinarian training is inadequate, said Wang. The primary focus of current legislation is on quality of animal production and supply, not on animal welfare.

Emerging Trends

Although enormous progress has been made, laboratory animal science across Asia, including China, is uneven and rudimentary. Factors influencing the quality of animal care and use include globalization, international exchange, economic development, and increased AAALAC-accredited institutions. Wang suggested that the most notable emerging trend in China is the shift toward alternatives to animal testing, which could lead to an overhaul of existing research paradigms in China. While the government is considering legislation that would eliminate animal testing of cosmetics, there appears to be no concerted effort to replace animal testing in other fields such as pharmaceutical, agrochemical, and chemical research.

SOUTH AMERICA: BRAZIL

Brazil is a diverse country with economic, religious, cultural, and social differences in addition to varied geographical areas with major cities and regions of the Amazon accessible only by boat or plane. Ekaterina Rivera, professor at the Biological Sciences Institute of the University of Goias in Brazil, said these differences lead to different approaches to animal research and animal regulations.

The southeastern part of Brazil has the largest cities and several established centers of excellence. In developing areas of Brazil, universities are new and just beginning to build laboratories and animal facilities. These universities and centers are working to catch up with the other regions. Rivera mentioned, for example, the National Institute of Research of the Amazon, which is mainly focused on nutrition studies using regional fruits. This institute has a very modern, specific pathogen-free animal house and is pursuing AAALAC accreditation.

Laws and Regulations

In 1985, a group of veterinarians and others created the National College on Animal Experiments (Colégio Brasileiro de Experimentação Animal [COBEA]) and issued the COBEA Ethical Principles based on the Council for International Organizations of Medical Sciences (CIOMS) principles. This was the starting point of laboratory animal science in Brazil. In 1986, COBEA became a member of the International Council for Laboratory Animal Science (ICLAS) and in 2009 its name was changed to the Brazilian Society of Laboratory Animal Science (SBCAL) to reflect its broader membership. SBCAL has been central to the promulgation of animal regulations in Brazil and has served as a model for other Latin American countries pursuing similar interests.

Until recently, Brazil had no regulations regarding the implementation of ethical committees; universities and institutions had nowhere to seek guidance or ask questions, Rivera said. Currently the primary Brazilian research body, the National Counsel of Technological and Scientific Development (CNPq), prohibits grants to be awarded to projects that have not passed an ethical review. This, Rivera said, was a critical step toward better science and animal care in Brazil as it caused a significant change in the approach scientists have toward laboratory animals.

The Arouca Law

In 1995, Sergio Arouca, a physician and a federal congressman, led a group of scientists in developing the first draft law governing the use of animals in research. Approved 13 years after it was drafted, the Arouca Law (Law n°11.794-October 8, 2008) regulates the use of vertebrate animals in research, teaching, and testing. The law also created the National Council for the Control of Animal Experimentation (CONCEA), which registers institutions, regulates experiments, and requires institutions to establish an ethics committee for the review of projects using animals. This law is similar to U.S. and EU laws in that it includes mention of the 3Rs, specifically the use of alternatives to animals and the avoidance of additional pain and distress. Rivera noted, however, that there is more emphasis on ethical committees in Brazil than elsewhere. The Arouca Law was subsequently regulated by Presidential Decree (n°6899/2009), providing the framework for CONCEA and its governing rules and functions.

Emerging Trends

Although relatively new, Rivera offered several comments on the effects of the laws and regulations over the past 10 years. Ethical committees have led to a "culture of care," changing the minds of those conducting research using animals. Brazil now has better animal houses, better trained personnel, and ultimately, better animals. In addition, veterinarians have a much larger role than before. There is a focus on alternatives to the use of animals and interdisciplinary meetings to advance animal science. In essence, the law helped set Brazil on the same path as its peers in other countries.

Some issues still need to be resolved, Rivera noted. For example, distance presents a significant challenge in a country the size of Brazil, affecting proper transportation of animals and access to supplies. The size of the country also affects the ability to inspect all of the laboratory animal houses. Rivera also noted that like any new law in Brazil, implementing the new animal law involves a good deal of bureaucracy.

Like Asia, the countries in South and Central America are diverse and at different stages in the development of their animal research regulations. The Brazilian law stimulated discussions across Latin America, and has served as a starting point for countries that still have no regulations. Currently, Brazil, Mexico, and Uruguay have specific laws on the use of animals in research. Argentina, Chile, Colombia, and Costa Rica have animal welfare laws with at least one provision relevant to laboratory animal science.

SUMMARY

Following the presentations, panelists and participants discussed the similarities and differences across countries and regions. Several speakers and participants noted there is a need to reduce bureaucracy that could hamper the progress of science, the value of increasing public confidence about the role of animals in research and the regulatory system, and the potential usefulness of developing appropriate metrics of success in balancing scientific quality, animal welfare, and public confidence (summarized in Box 2-4).

BOX 2-4
Summary of Session Points

Science Within the Regulatory Environment
- A primary role of research support services is to facilitate research by making compliance as seamless as possible for the investigator, while still assuring institutional compliance.
- The cost of regulation can be both a fiscal as well as a time cost for regulators and scientists (i.e., administrative burden).
- Bureaucracy is inevitable and may have a greater impact on animal research regulation enforcement in some countries more than others.

Animal Welfare and Scientific Quality
- Suggestions for improving both animal welfare and scientific quality:
  o Minimize non-experimental (e.g., environmental) confounding variables.
  o Improve the environment in which the animals are maintained to reduce experimental variability.
  o Reduce unnecessary duplication of studies.
  o Enhance the productivity of current animal models.

Public Confidence
- Public confidence in the regulatory process may increase if the scientific community helps to educate the public and politicians about
  o the nature of fundamental research,
  o how animal research contributes to science, and
  o animal research regulatory laws, policies, and requirements.
- Engagement by the scientific community in the regulatory and legislative process may result in more scientifically-based animal research regulations.
- Science might benefit from fostering relationships with alliance partners (e.g., patient groups, industry, animal welfare organizations) to bring a shared message to legislators.
- Mechanisms are in place for scientific organizations to comment on proposed regulations.
- Scientists in regions such as Asia and South America with less developed animal research regulations may have greater opportunities to participate in the process and educate lawmakers.

SOURCE: Individual panelists and participants.

3

Emerging Legal Trends
Impacting Animal Research

Unlike humans, animals cannot provide or deny consent for experimentation; therefore, animal protection must come from self-imposed rules or external government guidelines and regulations, said session chair Arthur Sussman. Against this background there is growing concern about the quality of animal-use enforcement and increasing public demands that there be new ways to address the interests or rights of animals. Panelists discussed the impact of current legal trends on the use of animals in research, specifically, emerging animal rights laws and the use of Freedom of Information requests at the national level and open-record laws or sunshine laws at the state level to gain access to information.

RIGHTS, REGULATORY SYSTEMS, AND REGULATION

Margaret Foster Riley, professor at the University of Virginia School of Law, suggested that a consideration of animal research regulations might extend beyond "regulation" to a consideration of the interactions among different forms of regulations, different systems that create and enforce regulations, and different uses of "rights." Rights involve many different social interactions. There are moral rights and legal rights. Rights are often not absolute and, in most legal systems, different rights may conflict with each other.

Recalling Blakemore's discussion of the cost-benefit analysis (Chapter 1), Foster Riley said that modern moral values can affect the way the law reflects people's perceptions of rights. Moral norms change over time, but not necessarily as a result of ethical or philosophical arguments. Philosophy

21

sets the foundation for cultural values, but context and psychology play significant roles in what is culturally acceptable. An individual's relationship with animals, for example, colors his/her general view on issues relating to animals.

Foster Riley said the perception of research using animals is also impacted by the perceived decline in the value of science in the United States. Foster Riley noted that while the Bayh-Dole Act[1] achieved its intended goal to facilitate the translation of basic science into clinical practice, it also resulted in a "loss of purity" for science. A public perception is that science now is done for profit and is subject to the influence of industry and special interest groups. This decline in the status of science has had a profound effect in law, noted Foster Riley.

Legal Rights, Property, and Legal Standing

In the public eye, legal rights are often a moral issue, with many believing that animals should not be viewed as property. For legal scholars, this is a crucial issue that determines what rights are at stake and who has legal standing to bring a lawsuit. It has been suggested that animals should be able to bring lawsuits themselves, in the same way that a child or a corporation can. Another approach is that people should have broader rights to bring lawsuits on behalf of animals.

Some of the most creative current litigation on animal research regulations, Foster Riley said, involves new uses of the False Claims Act, with lawsuits based on an improper use of public money. Under the National Institutes of Health (NIH) grants policy, if an investigator violates Public Health Service (PHS) policy on animal care or the Animal Welfare Act, the institution must return the NIH funds used in violation of grants policy. Moreover, an institution can be found in violation of government policy if it fails to report a problem once it is discovered. Under a "false claim" an argument could be made that the institution was claiming to be conducting research appropriately when indeed it was not, noted Foster Riley.

Trade law also can affect animal use. For example, Foster Riley suggested that at least some of the opposition to genetically modified organisms in Europe is based on trade protectionism concerns rather than public health and environmental concerns.

---

[1] The Bayh-Dole Act (Public Law 96-517, Patent and Trademark Act Amendments of 1980) created a uniform patent policy among the many federal agencies that fund research, enabling small businesses and non-profit organizations, including universities, to retain title to inventions made under federally funded research programs (NRC, 2006).

FREEDOM OF INFORMATION ACT

Margaret Snyder, Freedom of Information Act (FOIA) coordinator in the Office of Extramural Research at the NIH, noted that the NIH has a long history of transparency. As early as the 1950s, the NIH produced public booklets about its grant awards, listing the grants, institutions, investigators, and amount of funding. In 1998, the Computer Retrieval of Information on Scientific Projects (CRISP) database was launched, providing the same basic information as well as abstracts. The database could be searched for trends, techniques, specific projects, or particular investigators. Continuing this tradition of transparency, in September 2009, the NIH launched Research Portfolio Online Reporting Tools (RePORTER), the newest searchable database incorporating all of the information in the CRISP system as well as publications and patents.

One question that often comes up is how much the NIH spends on animal research. Although it is not possible to disaggregate the budgets to identify money spent on animal research for individual projects, it is possible to get a sense through Institutional Animal Care and Use Committee (IACUC) information. Snyder calculated that about 47 percent of NIH-funded grants have an animal research-based component. This number has been fairly steady over the past 10 years. Snyder noted, with animals being used in about 70 percent of the projects awarded funding through the National Institute of Neurological Disorders and Stroke.

FOIA Requests

Snyder gave a brief overview of FOIA (Box 3-1) and noted that for fiscal years 2006 through 2010, the NIH received 6,055 FOIA requests. Requesters are seeking information on a range of topics, including health information for themselves or a family member with a severe illness. The NIH also receives a considerable number of requests from individuals at academic and research institutions, some of whom are looking for a successful grant to model their application after, while others are seeking data for a policy or funding analysis for a scientific publication.

The NIH also receives FOIA requests from animal advocates. Snyder noted that the number of requests to 6 institutes with neuroscience activities[2] from advocates peaked in 2008; of the 70 FOIA requests from animal advocates across 27 institutes and centers, 35 (50 percent) were for

---

[2] The six institutes highlighted by Snyder are the following: National Institute on Alcohol Abuse and Alcoholism (NIAAA), National Institute on Deafness and Other Communication Disorders (NIDCD), National Institute on Drug Abuse (NIDA), National Eye Institute (NEI), National Institute of Mental Health (NIMH), and National Institute of Neurological Disorders and Stroke (NINDS).

BOX 3-1
Freedom of Information Act (FOIA) Overview

History
- 1966—FOIA signed into law by President Johnson.
- 1996—Electronic Freedom of Information Act (E-FOIA) signed by President Clinton; the Internet and electronic distribution should be used as a means to improve public access to records.
- 2007—Open Government Act signed by President G. W. Bush imposed consequences for agency non-compliance with new FOIA 20-day response time.
- 2009—Attorney General Holder, at the direction of President Obama, issued a Memorandum on a "New Era of Open Government," stressing transparency.

Provisions
- Provides the statutory right of a person or organization to obtain U.S. government information:
  - Including that from federal agencies.
  - Excluding the personal staff of the President, the Congress, or the courts.

Exemptions
- Nine exemptions are intended to protect privacy, financial information, and personal security.
- Exemptions most often used by the National Institutes of Health (NIH):
  - Exemption 4: "Trade secrets and commercial or financial information."
  - Exemption 5: Interagency or intra-agency memorandums and incorporates; certain privileges (e.g., deliberative process, attorney–client communications, and attorney work projects).
  - Exemption 6: Certain information if disclosure would constitute clearly unwarranted invasion of personal privacy (e.g., personnel actions).
  - Exemption 7: Protects certain information in law enforcement files.
- The responding agency may not consider the identity of the requester or what the requester intends to do with records when deciding whether to release or withhold records.

SOURCE: Snyder presentation.

---

these 6 institutes. Snyder noted that overall requests are not only increasing in number, but in magnitude. Of particular interest to animal rights advocates is information regarding research using non-human primates and information about institutions found in non-compliance. This second category can include institutions failing to report deficiencies or reports of actions by researchers in violation of either the Animal Welfare Act or PHS policy.

FOIA Exemptions

A request by researchers to withhold information must be targeted, and any assertion of harm from disclosure must be very specific. Only the NIH FOIA officer can decide to withhold information. A written record then becomes part of the file that will be used in the event of an appeal or litigation.

Snyder pointed out that even though the NIH has increased transparency though RePORTER, CRISP, and other venues, the number of FOIA requests is not declining. Of concern, she said, is that the NIH relies heavily on institutional self-monitoring and self-reporting. The NIH works closely with grantee institutions and engages them in an interactive, collaborative exchange of information to correct any deficiencies. The success of this process depends on maintaining a confidential, collegial, interactive relationship, and that relationship is now in jeopardy. The NIH is finding it more difficult to protect information. There are no more blanket or categorical exemptions; arguing the need for protection involves case-by-case, line-by-line review. She did note that the NIH perspective is that unfunded grants are protected as intellectual property.

The NIH releases names of researchers because that information is already public in the RePORTER database, Snyder said. In some cases, such as in non-compliance reports, she asks the FOIA requester if he/she will allow the NIH to redact the names of secondary individuals (i.e., the institutional official, chair of the IACUC, veterinarian, and investigator are disclosed and all others are redacted). In most cases the requester agrees.

STATE SUNSHINE LAWS

Richard Cupp, professor of law at Pepperdine University School of Law, explained that U.S. political power is shared between the federal government and the states. Each state has an open-records law, or "sunshine law," that is often very similar to the federal FOIA.

There are limitations on the type of information that can be obtained through state sunshine laws. Most of the disclosure obligations and exceptions under state sunshine laws are the same as FOIA. Possession dictates whether FOIA or state sunshine laws apply. If something is in the possession of a federal agency (e.g., NIH), a FOIA request is made; if something is in the possession of a state institution (e.g., the University of California, Los Angeles [UCLA]), that state's document release laws will apply. Some documents may be in the possession of both a federal agency and a state entity and there may be both a FOIA request and a state sunshine request. The 50 states have 50 different laws dealing with disclosing information, Cupp said. While they are generally overlapping, there are some distinctions.[3] California, for example, has a general exemption clause that allows an entity to withhold information even if there is not a specific exemption. The entity must demonstrate that on the facts of the particular case, the public interest served by not disclosing the record clearly outweighs the public interest served by disclosing the record. The state organization has a strong burden of proof.

Misuse of Sunshine Laws

Cupp noted that transparency is important to a democratic government and most of the information obtained under sunshine laws is used for its intended purpose of supporting democratic discussion and the democratic process. Some information, however, has been obtained with the intent to threaten and harass researchers and institutions.

Cupp shared a recent case at UCLA as an illustration of how sunshine laws have been used by U.S. activists. Two UCLA researchers received envelopes containing what an animal rights group press release described as "a dangerous present at their home." They also received other threats, including "we are watching you and we know about your wife" and "we have been following you for the past 6 months all over campus, to your car, visits to [a local grocery store], to movies, and to social gatherings." There is a long history of incidents similar to this at UCLA, including firebombing a researcher's car, use of Molotov cocktails to firebomb researchers' homes, and flooding of a researcher's home with a garden hose. Researchers have also received razor blades that were allegedly covered in HIV-infected blood.

During this time period, Cupp noted that UCLA and animal rights groups have battled over how much information about research is required to be disclosed under California's sunshine law. In one case that has been active for a number of years, the judge overseeing an appellate court review stated that there is a "causal nexus between [UCLA's] disclosure of animal research records and subsequent attacks on the researchers identified in such records after they are disseminated to the public." In another court ruling that was upheld on appeal, the judge found that releasing some of the documents sought by animal rights activists "would result in a significant and specific risk of unlawful intimidation and physical harm to researchers and to their families."

---

[3] The website http://www.sunshinereview.org provides information on all 50 states' sunshine laws.

Transparency

Cupp noted that greater transparency would not necessarily result in fewer FOIA or sunshine requests. Increased public access to documents would reduce the amount of work on the part of those tasked with responding to these requests. But would that be a good balance of risk and utility for an institution, Cupp asked?

Cupp suggested exercising care in writing documents and communications to avoid unnecessary disclosure of sensitive information, or information that may be misinterpreted. For example, researchers could keep e-mails related to research short and on point. Cupp also suggested that personal e-mail accounts not be used when writing about research as all personal e-mails might become subject to FOIA requests or state sunshine law requests. Cupp suggested that researchers avoid jokes or sarcasm that could be misunderstood.

He also encouraged institutions to make public and strong expressions of support for scientific research involving animals. With the encouragement of faculty members, the UCLA administration in 2007 began making strong press statements backing the research work of the university.

4

Animals in Neuroscience Research

Current and new regulations, including requirements to implement the 3Rs (replacement, refinement, and reduction), along with public desire to reduce the number of animals used, could potentially impact the speed and quality of biomedical research, noted Roberto Caminiti, professor of physiology at Sapienza University of Rome and session chair. Panelists discussed the role of animals in neuroscience research, benefits and costs (administrative, economic, social, animal welfare), mechanisms to maintain public confidence, and the impact of the laws, regulations, and policies on animal-based research in neuroscience (key points are summarized at the end of the chapter in Box 4-2).

RODENTS IN NEUROSCIENCE RESEARCH

Rodents are the dominant mammalian animal species used in neuroscience research, said Bill Yates, professor of otolaryngology and neuroscience at the University of Pittsburgh, but the Animal Welfare Act excludes mice and rats, so the exact number used in the United States is not available. The number of higher animals used is known because the U.S. Department of Agriculture (USDA) requires research institutions to submit an annual report of the number of animals used. The use of most animal species tracked by the USDA has declined over the past decades (Figure 4-1). Only the use of non-human primates has increased slightly. Yates noted that during this time, National Institutes of Health (NIH) grant funding has increased tremendously, suggesting that if more animal research is being done, it must be in species such as rodents, which are not regulated by the Animal

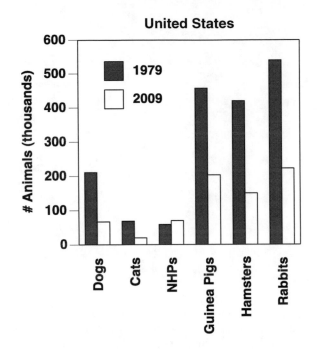

FIGURE 4-1 U.S. Department of Agriculture (USDA)-tracked animal use data for the United States, 1979 and 2009. (Data for rats, mice, birds, and cold-blooded vertebrates are not tracked.) NHP = non-human primate.
SOURCE: Yates presentation citing USDA Annual Reports.

Welfare Act. The UK Home Office tracks the number of procedures (not the number of animals used) and does include rodents. Over the past 20 years, the use of all animal species except mice has decreased (Figure 4-2).

Increased Use of Rodents

Prior to the mid-1980s, cats were popular research animals for classical neurophysiological procedures because they could withstand the extensive surgeries required, were large enough to accommodate bulky instrumentation, and were inexpensive models. However, in the mid-1980s, new regulations substantially increased the economic cost and administrative burden of feline models. In addition, public opinion shifted against the use of companion animals in research.

Miniaturization of instrumentation has allowed rodents to serve as replacements for felines in some studies. Refinement of techniques such

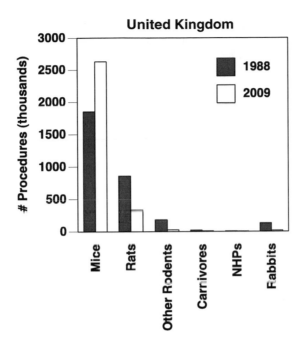

FIGURE 4-2 UK Home Office–tracked animal procedure data, 1988 and 2009.
NHP = non-human primate.
SOURCE: Yates presentation citing UK Home Office Web site.

as chronic recording techniques enables the study of a single animal over a prolonged period of time, leading to greater data collection from single animals. This results in fewer animals needed per study. Thus, the use of non-human primates, which can be trained for more elaborate tasks than cats, has become more economically feasible. Refinement, Yates pointed out, does not always lead to use of a lower species.

Transgenic Mouse Models

The most significant contributor to the increased use of rodents in biomedical research has been the development of transgenic mouse models. In the late 1980s, Capecchi, Evans, and Smithies developed principles for introducing specific gene modifications in mice by the use of embryonic stem cells leading to the development of the first knockout mouse. Today, human genes can be inserted into a mouse or overexpress a particular gene.

Through breeding, it is possible to obtain a line of animals that expresses a new phenotype. Most procedures now are done using transgenic animals. Data suggest that transgenic mice likely account for two-thirds or more of the mice, and more than half of the mammals used in biomedical research.

Use of transgenic animals has allowed neuroscientists to decipher the function of particular genes and to create disease models, Yates said. Knockout models have been used in the study of Alzheimer's disease, for example, and have been critical in understanding the neural basis of learning and memory.

Use of transgenic mouse models does have limitations, Yates noted. Genetic diseases involving multiple genes can be difficult to model in transgenic animals. In addition, some genetic diseases have different phenotypes in mice and humans. For example, transgenic models of Parkinson's disease often do not exhibit the same neural degeneration observed in humans. In addition, compensation for the gene manipulation during development can lead to false conclusions about the role of particular genes.

Rodents Versus Higher Mammals

Even considering the limitations, transgenic animals and rodents in general have provided a significant boost to biomedical research. But are they the ideal research model? Yates highlighted some of the advantages and disadvantages of using rodents (Box 4-1).

Expanding Transgenic Technology to Other Species

The technology is now available to create other transgenic species. Zinc-finger nuclease technology has allowed the creation of knockout rats, and theoretically, the technique could work for inactivating genes in any species, including humans. As more types of transgenic animals become available, the balance of species used in biomedical research may shift, Yates noted.

THE ROLE OF NON-HUMAN PRIMATE MODELS IN NEUROSCIENCE

Roger Lemon, Sobell Chair of Neurophysiology at the University College London Institute of Neurology, showed data from the UK Home Office spanning from 1995 through 2010 that indicates a gradual increase in the use of old-world monkeys (primarily macaques) and a gradual decrease in the use of new-world monkeys (mainly marmosets). Overall, non-human primates were used in a very small percentage, less than 0.1 percent, of the total number of procedures involving animals in the United Kingdom.

BOX 4-1
Are Rodents Ideal Research Models?

Advantages
- Rodents typically live <2 years, which facilitates aging studies.
- The small size of rodents allows many animals to be maintained in a limited space.
- In some cases, rodents are better models of human diseases than other animal models.

Disadvantages
- Rapid aging can confound repeated measures over time in the same subject.
- The small size of rodents causes constraints on manipulations and measurements.
- Some human disease conditions cannot be mimicked in rodents.

Other Concerns
- Some findings in rodents might be unique to rodents.
- Rodents lack the ability to vomit, so carnivores remain a better model animal to study emesis and conditions that elicit emesis (e.g., motion sickness).
- Rodents lack respiratory responses such as coughing and sneezing; carnivores remain the most appropriate model for studying cough-suppressing drugs.

SOURCE: Yates presentation.

The majorities, about 81 percent, were involved in applied research (e.g., toxicological tests), often due to a statutory requirement for testing of new drugs in a non-human primate model before entering human clinical trials.

The Case for Non-Human Primate Models

Regulatory Opinion

Recital 17 of European Union (EU) Directive 2010/63 states that "the use of non-human primates in scientific procedures is still necessary in biomedical research," and that "the use of non-human primates should be permitted only in those biomedical areas essential for the benefit of human beings, for which no other alternative replacement methods are yet available." Recital 13 states that the methods selected should "require the use of species with the lowest capacity to experience pain, suffering, distress or lasting harm, that are optimal for extrapolation into the target species." In essence, then, Lemon said, both recitals urge that non-human primates be used only in those areas that are likely to be of ultimate benefit for humans.

Articles 5 and 8 of the directive state that non-human primates can be used for basic research. Much of the basic research work in the United Kingdom using non-human primates involves understanding the role of the prefrontal cortex, which may inform progress toward treatment of human neurological and psychiatric disorders of the frontal lobe. The rodent, Lemon noted, is not a particularly good model for studies of higher-level cognitive processes as it lacks the cortical complexity of the human brain.

Independent Policy Reports

Several reports have outlined the scientific case for continued use of non-human primates in biomedical research, including the Weatherall Report in 2006 (MRC, 2006) and the EU Scientific Committee on Health and Environmental Risks (SCHER) report in 2009, both of which identified neuroscience in particular as an area where evidence supports the use of primates (MRC, 2006; SCHER, 2009). A 2004 report from the Academy of Medical Sciences highlighted the need to promote translation of basic science into clinical practice to improve neurorehabilitation, including better therapies for rehabilitation of hand function. This is very clinically relevant, Lemon noted, as in the United Kingdom there are 100,000 new cases of stroke every year and about half of these patients will have some form of serious hand disability. Loss of hand function is also associated with spinal injury (800 new cases per year in the United Kingdom) and cerebral palsy (1,800 new cases each year in the United Kingdom). Lemon also alerted participants to the Bateson report, a retrospective survey of research in the United Kingdom using non-human primates, which was expected to be released the same week as the Institute of Medicine workshop (MRC, 2011).[1]

Neuroscience Research

Lemon noted that a review by Courtine and colleagues (2007) concluded that there are "crucial differences in the organization of the motor system and behaviors among rodents, non-human primates, and humans" and that "studies in non-human primates are critical for the translation of some potential interventions to treat spinal cord injury in humans."

There are major differences in the organization of the corticospinal system across species, Lemon said. Examples include the size and numbers of fibers involved; the trajectory that neurons follow within the spinal cord; the extent to which they reach within the spinal cord; and how

---

[1] The report of the independent panel chaired by Patrick Bateson was released on July 27, 2011. The findings were not discussed at the workshop because the report was not publicly available until the second day of the workshop.

they terminate within the spinal gray matter. In primates, the extent of cortico-motoneuronal connections correlates with dexterity, and all dexterous primates that use tools in the wild have highly developed cortico-motoneuronal connections (Lemon, 2008).

Lemon highlighted the work of Schwab and others as an example of how studies in non-human primates can lead to clinical trials. In the late 1980s, Schwab discovered that axons on the spinal cord contained a protein that inhibits the growth of neurons, which he subsequently named Nogo, for "NO GrOwth" (Schnell and Schwab, 1990). In vitro studies showed that neuron growth in culture was strongly suppressed by the myelin inhibitory factor Nogo. Over the next 15 years, Schwab conducted studies in mice and rats to characterize the properties and mechanism of action of Nogo, and developed a means of neutralizing it with antibody (anti-Nogo). Only after this extensive fundamental research, Lemon said, did Schwab decide it was necessary to move to a primate model. The first primate study assessed anti-Nogo as a potential treatment for spinal cord injury (Freund et al., 2006). Non-human primates with untreated spinal lesions permanently lost the ability to make hand movements smoothly, efficiently, and accurately; non-human primates treated with the Nogo-specific antibody largely recovered their ability to make dexterous movements. Lemon stressed that it would be very difficult to assess the impact of spinal lesions on hand function in a rodent. As a result of this successful study in macaques, a Phase I clinical trial in humans has been completed and a Phase II trial began in 2010 (Zörner and Schwab, 2010).

The Future of Non-Human Primate Research

Non-human primate research will continue to be needed, especially research directed at lifelong conditions such as neurodegenerative diseases and psychiatric disorders, Lemon opined. Studies using non-human primates complement other data collection approaches, such as in vitro studies, in silico modeling, human brain imaging, and parallel investigations in rodents. The number of animals used will be relatively low; however, long-term study of a single primate can involve a significant number of independent assessments, resulting in reliable statistical answers from relatively small numbers of animals.

There is a very positive culture of non-human primate care in the United Kingdom, Lemon said. The UK National Center for the 3Rs has played an important role in training and raising the standards of care and knowledge among those working with primates, including technicians, animal care staff, postdoctoral fellows, and principal investigators.

Lemon noted cost, regulatory burden, and training as issues impacting the use of non-human primates in neuroscience research in the United

Kingdom. Cost is a significant obstacle for UK researchers. The purchase cost for a single purpose-bred macaque (excluding taxes) is about £20,000 (more than $30,000) and per diem costs for housing and care range from £50 to £70 per macaque per day (about $80 to $110 per day). Some of the cost stems from increasing standards of welfare that are required and additional security needs. On the upside, the high cost effectively ensures that no trivial or unnecessary work is done in non-human primates. The downside, however, is that high economic costs threaten serious non-human primate research in the United Kingdom. A participant commented that similar financial challenges face non-human primate researchers in other countries. Without additional investment in infrastructure, Lemon observed, centers that are using non-human primates may find it difficult to compete with other types of research in the long term.

The possible reclassification of "moderate procedures" involved in long-term neuroscience studies as "severe" is another problem facing EU researchers. Lemon suggested that reclassification may lead to large restrictions in the types of neuroscience research that can be conducted on non-human primates. Finally, training is important for the long-term future of non-human primate research. Lemon suggested that the perceived difficulty of conducting research with non-human primates may negatively affect the ability to attract the best young scientists to the field.

ETHICAL AND PRACTICAL DILEMMAS OF RESEARCH WITH NON-HUMAN PRIMATES

Basic Versus Applied Research

Stuart Zola, director, Yerkes National Primate Research Center at Emory University, noted that the definitions of basic (or fundamental) research and applied (or translational) research are not necessarily clear. In the early 1600s, Sir Francis Bacon divided research into *experimenta lucifera,* experiments shedding light, and *experimenta fructifera,* experiments yielding fruit. The distinction between basic and applied research is relevant to the use and regulation of animals in research. Biomedical ethics committees and Institutional Animal Care and Use Committees (IACUCs), for example, must consider the potential benefits of the proposed research for humans and animals. In addition, animal rights groups are often concerned that basic or fundamental research using animals has no immediate application to humans.

In practice, it can be difficult to distinguish to which domain an activity clearly belongs. For example, an experiment that involves the development of a behavioral task in non-human primates to assess functions of the hippocampus would seem to be very basic research. However, there is clear application of the knowledge in terms of diagnostics or interventions with respect to a wide range of clinical diseases and conditions. Zola noted that what may look like basic research may have very clear applications. Basic research was critical to the development of medical breakthroughs such as coronary bypass surgery and magnetic resonance imaging (MRI), for example. In the United States and the United Kingdom, Zola noted, the focus is now "translational research," bringing together basic scientists and clinicians to develop the best and most effective treatments and interventions.

The "Justification Rule"

An issue of concern for scientists is the idea that some clear applied benefits should come from the research itself, whether it is a diagnosis, treatment, prevention, or some other benefit to humans or animals. This "justification rule" is espoused in EU Directive 2010/63 (Para 17) which states that non-human primate research "should be permitted only in those biomedical areas essential for the benefit of human beings, for which no other alternative replacement methods are yet available" or when basic research is carried out in relation to potentially life-threatening conditions in humans or in relation to cases having a substantial impact on a person's day-to-day functioning (i.e., debilitating conditions). Zola opined, however, that this need for justification is based on two presumptions that are incorrect: first, that there are clear distinctions between basic research and applied research, and second, that it is possible to predict direct benefits to humans or animals that result from research using animals. Instead, Zola said, we can recognize that the discovery of fundamental knowledge has value in its own right. This is not an "anything goes" approach, Zola stressed, but an approach of basing choices on science and value, and not on semantics and arbitrary distinctions.

Challenges to Non-Human Primate Research

Advances in technologies related to genomics, behavior, imaging, and microbiology/immunology are offering new avenues for non-human primate researchers to develop therapies, interventions, and diagnoses (Figure 4-3). Zola offered several examples of challenges and welfare concerns facing researchers related to some of these new technologies. Positron emission tomography (PET) imaging can be used to conduct brain imaging while the animal is engaged in cognitive tasks, very much the same way it would be done with humans, Zola noted. However, a number of concerns are associated with PET imaging of non-human primates. First, the animal is awake, raising questions about stress, not unlike the stress many humans feel when inside a much longer MRI tube.

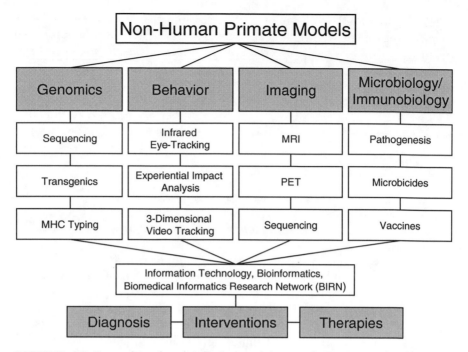

FIGURE 4-3 Examples of technologies used in translating research with non-human primates to human applications.
NOTE: MHC = major histocompatibility complex; MRI = magnetic resonance imaging; PET = positron emission tomography.
SOURCE: Zola presentation.

The use of PET imaging has recently been used to show the localization of simian immunodeficiency virus (SIV) in individual macaques. The ability to track the virus in the body is revolutionary, Zola noted, and will aid development of HIV vaccines and interventions. However, this requires a lot of animals, and the particular macaque species used in this study, the sooty mangabey, is an endangered species. This use is relevant as the sooty mangabey is one of the species from which the mutation from SIV jumped to humans. Research in this species could help answer many questions about immunity to SIV, but invasive research is prohibited because the sooty mangabey is endangered.

A rapidly advancing area is the development of transgenic non-human primate models of inherited neurodegenerative diseases. Recently, researchers produced the first transgenic non-human primates that express the Huntington's disease gene; the animals exhibit many of the defining signs of Huntington's disease (Yang et al., 2008). Animals also have been developed that carry risk factor genes for Alzheimer's disease. Longitudinal studies of the animals are ongoing, including gene expression studies, MRI, and cognitive behavioral evaluation. This is a remarkable new era in the study of disease, Zola said, but there are ethical concerns about inducing disease in non-human primates. In addition, such studies require a large number of animals.

Overall, the most significant challenge is infrastructure. The lack of resources, space, animals, and funding is outweighing the ability to do the research. The precarious balance between science and infrastructure is really an ethical concern, Zola concluded, as the inability to do the science will lead to lives lost in the end.

ADMINISTRATIVE AND ECONOMIC COSTS

Charles J. Heckman, professor at Northwestern University Feinberg School of Medicine, offered his perspective on the regulatory burden from the viewpoint of an IACUC chair. Across Northwestern's two campuses, there are a total of about 16,000 cages of mice at any given time, approximately 500 cages of rats, and a modest number of larger animals. Approximately 200 principal investigators are involved in animal research. The university is AAALAC-accredited, and receives NIH funding totaling around $300 million per year, about half of which is probably associated with animal research, Heckman estimated.

Consequences of Regulations for a Large Research University

Protocol Review

An animal program the size of Northwestern's leads to a large amount of work for the IACUC and the administration. At any one time, there are between 900 and 1,000 animal protocols, Heckman said, with about 200 new protocols each year. In the U.S. system, protocols must undergo a full review and renewal every 3 years, meaning about 50 to 100 de novo reviews each year. There are also 250 to 300 personnel addendums each year, some covering multiple people.

Facility Inspections and Reports

As mandated by the Public Health Service (PHS) Policy and the Animal Welfare Act, all animal housing spaces must be inspected twice per year. The inspection teams include several IACUC members, an IACUC staff person, and sometimes a safety staff member. It takes a minimum of 5 teams a minimum of 2 hours each to inspect a facility. The laboratory spaces of the 200 investigators conducting surgical procedures must also be inspected twice per year and the semi-annual reports consume a modest amount of time and effort as well.

IACUC Staff Personnel and Volunteer Members

The total IACUC staff handling this administrative load at Northwestern is 7 full-time positions: a director, 3 IACUC program assistants, an administrative assistant, and 2 people responsible for postapproval monitoring. There are 23 volunteer IACUC members: 3 veterinarians, 3 community members, 2 members from the Office of Research Safety, and 15 principal investigators from various departments with significant animal programs. Heckman estimated his own effort as chair at around 15 to 20 percent, and noted that it is difficult to find a sufficient percentage of effort to sustain his scientific work while serving as IACUC chair. Based on the review burden, Heckman estimated that a typical committee member needs at least a couple of hours per week to review protocols and attends a 2- to 3-hour committee meeting each month. A subcommittee of the IACUC is also devoted to reviewing medical records, which takes 2 to 3 hours per week. Each lab working with a USDA-covered species also must conduct a monthly self-audit of at least some of their medical records and report the audit findings to the IACUC.

Investigators and Laboratory Members

Protocol preparation takes a significant amount of time. Each individual protocol is approximately 30 to 40 pages, taking at least 2 hours per protocol to draft, followed by several rounds of revisions and review. Most large laboratories with four or more protocols will usually have a laboratory manager who dedicates at least a third of his or her time to managing the protocols. The principal investigator is ultimately responsible for ensuring that fellows, students, and staff understand the importance of the process, and adhere to the approved protocols. This is not just paperwork, Heckman stressed.

BOX 4-2
Summary of Session Points

Basic vs. Applied Research
- Arbitrarily separating research into "basic" and "applied" categories could be harmful if used to determine the types of research that can or cannot be conducted:
  o The line between basic and applied research is often blurred.
  o The discovery of fundamental knowledge has value in its own right.
  o Some basic research may have clear application toward development of treatments for nervous system diseases and disorders.
- It is not always possible to predict specific benefits to humans or animals that result from research using animals.

Animals in Neuroscience
- Rodents are the dominant mammalian species used in neuroscience research:
  o Miniaturization of instrumentation has allowed rodents to replace larger animals.
  o Development of transgenic mouse models has significantly increased the use of rodents.
- Refinement does not always lead to use of a species with lower cognitive abilities, such as rodents.
  o Refinement of techniques can result in a decreased number of animals required for study, thus making the use of species of higher cognitive abilities, such as non-human primates, more economically feasible.
- Non-human primates continue to be used in biomedical research, including neuroscience:
  o Studies in non-human primates can lead to human clinical trials.
  o Non-human primate studies can complement in vitro studies, in silico modeling, human brain imaging, and parallel investigations in rodents.
  o Long-term study of a single primate may involve numerous independent assessments, resulting in reliable statistical answers from a relatively small number of animals.
  o Difficulties in using non-human primates in research include costs, regulatory burden, and attracting talent to the field in the face of mounting challenges.
  o The potential reclassification of "moderate procedures" in long-term neuroscience studies as "severe" may impact the use of non-human primates.

Challenges Associated with New Regulations
- Increased recordkeeping requirements.
- Increased cost of conducting biomedical research without direct evidence that increased regulations result in improved animal welfare.
- Conflicting regulations from multiple agencies with multiple sets of rules.
- Compliance can require significant time and effort from dedicated IACUC staff, IACUC volunteer members, principal investigators, and laboratory staff, for writing and review of protocols and addendums, animal facility and laboratory inspections, monitoring, and reports. Training and accreditation activities can consume resources.

SOURCE: Individual panelists and participants.

5

Advancing the 3Rs in Neuroscience Research

The 3Rs (replacement, refinement, and reduction) play an increasingly important role in animal research regulations. As previously described (see Chapter 2), the revised European Union (EU) Directive includes a formal introduction of the 3Rs as guiding principles. In addition, both U.S. and Chinese regulations call for incorporation of the principles of the 3Rs in experimental design. This session explored examples of how the 3Rs are implemented in two fields of neuroscience research, spinal cord injury and epilepsy. Speakers also discussed how systematic reviews could be applied to preclinical research to help advance the 3Rs.

Sue Barnett, professor of cellular neuroscience at the University of Glasgow, opened this session with a brief introduction to the 3Rs, the framework for the humane use of animals in research first articulated by Russell and Burch in 1959 (Box 5-1). (Session points are summarized at the end of the chapter in Box 5-2.)

REPLACEMENT CASE EXAMPLE:
SPINAL CORD INJURY MODELS

Barnett described an example of a replacement strategy she is developing for spinal cord injury research. Clinical strategies have primarily been palliative care, including drugs (e.g., steroids) to dampen the immune response during the acute phase, advanced rehabilitation strategies (e.g., physiotherapy), and neural prostheses (e.g., functional electric stimulation [FES]).

The most common causes of spinal cord injury are motor accidents (50 percent), falls (24 percent), and sports (9 percent). Spinal cord injury

---

BOX 5-1
Replacement, Refinement, and Reduction

Replacement
Methods to **avoid or replace the use of animals** in areas where they otherwise would have been used, including
- using non-animal alternatives such as human volunteers, computer models, and in vitro techniques.
- using animals of lower neurophysiologic sensitivities such as invertebrates.

Refinement
Improvement to scientific procedures and husbandry that **minimize pain, suffering, distress, or lasting harm and/or improve animal welfare**, including
- improved procedures (e.g., surgery).
- improved anesthesia.
- improved housing and husbandry.
- better welfare assessment.

Reduction
Methods that **minimize the number of animals used** (or maximizing information gained from a given number of animals), including
- good environmental design and statistical analysis.
- tissue sharing.
- imaging.

SOURCES: Barnett presentation; Russell and Burch (1959).

---

is a complex event that begins within minutes of the mechanical injury and progressively worsens over the subsequent weeks to months.

Repair Strategies

After an injury, formation of glial scars inhibit central nervous system repair by creating both physical (e.g., cyst) and biochemical (e.g., inhibitory signals) barriers to axonal growth. The goal of any repair strategy is to fill any cysts, maintain glial/neuronal survival, limit scar formation, promote axonal regeneration, and make functional reconnections. Using animal models, researchers are studying injecting growth factors, blocking inhibitory signals (e.g., anti-Nogo [described by Lemon]), transplanting cells, bridging the gap using biodegradable scaffolds to align the axons, and promoting plasticity/sprouting of any remaining intact fibers. No one treatment alone is capable of repairing the spinal cord, Barnett noted. Current thought is that a combination of strategies will be required.

Three main laboratory strategies are currently used to treat a damaged spinal cord. The first, *neuroprotection,* is to protect what is left and minimize further damage. Second, especially for incomplete injuries, the strategy includes *remyelination* or making the most of what remains. *Repair* is the third strategy, which includes restoring communication, axonal regeneration, and reconnection, often by cell transplantation or pharmacological intervention.

Spinal Cord Research in Animals

Spinal cord research in humans is difficult and in some cases impossible. There is no ability to biopsy tissue, imaging is limited, and studies cannot be done on large groups of people with similar pathology. The only way to investigate spinal cord injury, Barnett said, is to use animal models or primary cells from animal tissue.

An example of an animal model of a spinal cord lesion is a wire knife lesion, generated by inserting the knife into the dorsal column and pulling up a piece of tissue. Barnett noted that this method is clean, accurate, and consistent, resulting in a cavity and glial scarring that mimics human spinal cord injury. By tracing regenerating axons using fluorescent labeling techniques, Barnett has observed that while many axons enter and fill the lesion site, they have limited ability to grow through the lesion, and few exit and find their target. This, Barnett explained, is the major problem with many of the spinal cord injury repair therapies.

One aspect of spinal cord injury that researchers want to mimic is the glial scar. A useful model would have a lesion surrounded by reactive astrocytes that express molecules of interest; axons would be inhibited from entering or exiting the scar and would become demyelinated; and there would be activated microglia.

Several disadvantages to rat models of spinal cord injury include the need for large numbers of animals, the severity of the procedure, and the distress and discomfort to the animals, Barnett said. Additionally, there is a long time frame for results and the experiments are expensive and time consuming. To address this, Barnett is working to replace animals in her experiments.

Replacing Animals with Cell Culture

Barnett described her in vitro model of spinal cord injury in which disassociated embryonic spinal cord cells from rats are layered on top of an astrocyte monolayer derived from embryonic tissue (Sorensen et al., 2008). Growth in culture over time leads to complex axonal/glial interactions resulting in myelinated neurons. This system allows for the study of contact between astrocytes and how they communicate with the axons, which is necessary for understanding these problems in spinal cord injury. Barnett and colleagues next induced lesions in the cell culture by cutting with a scalpel to studying axon density and myelination adjacent to the lesion and cell growth into the damaged area. To validate the model, Barnett has studied several molecules previously tested in vivo to see if they could promote outgrowth or repair in the vitro system.

Overall, the findings from the in vitro model of spinal cord injury correlate with in vivo findings, including the formation of features typical of a glial scar, neurites that do not cross the boundary of the scar, and myelination and neurite density that is decreased adjacent to the lesion. The cells in culture respond to reagents that have been reported to promote axonal growth in rat models of spinal cord injury. This model also could be used to prescreen combinations of biological and pharmacological agents for potential therapy for repair of spinal cord injury. Barnett noted that getting the model published so others can become aware of it has been successful, but also challenging (Boomkamp et al., 2012).

REFINEMENT AND REDUCTION CASE EXAMPLE:
EPILEPSY MODELS

Gavin Woodhall, reader in neuropharmacology at Aston University, discussed refinement and reduction strategies, using his work in epilepsy research as an example. Refinement can improve research findings, he noted, and often results in reduction as a "byproduct." Simple refinements can have significant effects on the study results. Enriching the cage environment, for example, by adding a few tunnels or a bit of nesting material to a rat cage, improves the neurological development of rats. Rats reared in an environment that contains no enrichment show different somatic mechanisms of memory than rats that have been reared in an enriched environment. Studies have also shown that cross-fostering to equalize litter sizes impairs cortical neuronal network function. Other examples of refinement include substitution of non-invasive approaches for more invasive ones; use of analgesia preoperatively, not just postoperatively; habituating animals to procedures, such tail-vein blood sampling, so that they are less stressed; and reducing the severity of protocols.

Animal Models of Epilepsy

In the United Kingdom, 450,000 people, or 0.5 to 1 percent of the population, suffer from epilepsy, with approximately 30,000 new cases diagnosed each year. One-third of patients do not respond to any of the currently available drugs and 20 to 30 percent do not improve with surgery. In the developing world, 60 to 90 percent of epilepsy is undiagnosed or untreated.

A variety of in vivo and in vitro models of epilepsy exist, including spontaneously induced epileptic mouse strains, chemically or physically induced models, and cultured neurons. Woodhall's research relies on a long-established technique called lithium-pilocarpine epileptogenesis, which uses a chemical insult to provoke development of epilepsy over an extended period of time and results in a chronic epileptic syndrome in an animal. Brain slices are then obtained from the animals for testing. After injection of the drugs, the rodent goes into acute status epilepticus defined as continuous seizures with very short gaps in between. In many laboratories, this phase is allowed to continue anywhere from 90 minutes to 6 or 7 hours, Woodhall said. Seizures are then arrested with a sedative. The animal enters a quiescent period that lasts 1 or 2 weeks before they begin to exhibit spontaneous recurrent seizures. A conservative estimate of mortality from this approach is 5 to 50 percent; however, in some laboratories mortality rates are more than 80 percent, Woodhall noted. This extremely high mortality rate prompted Woodhall to focus on how this model could be refined and survival improved.

Questions persist as to whether these models are good models of temporal lobe epilepsy, or of epilepsy in general, and whether the pathology is similar to that seen in humans. There are also concerns about reproducibility, as measurement of key indicators can be highly variable. For example, γ-amino butyric acid (GABA)–mediated levels, an indicator of inhibitory action, in this model have been shown to decrease, increase, or remain unchanged. In addition, Sloviter (2005) showed that when animals are allowed to remain in acute status epilepticus for 6 or 7 hours, large areas of hemorrhage and damage were visible in brain slices. This raises questions about the seizures the model elicits, specifically whether these seizures as a result of gross global damage are a true model of human epilepsy, said Woodhall.

Refinement

Woodhall raised several questions regarding refinement of the current epilepsy model: whether the severity of this approach can be reduced; whether acute status epilepticus can be avoided altogether; whether more "ethical value" can be gained from the model; and whether other approaches could be used.

Seizure activity feeds from the cortex, through the basal ganglia, and back into the cortex, to create a positive feedback loop during epileptogenesis. Seizures then become uncontrolled and spread to the brainstem, killing the animal, Woodhall explained. Use of the central muscle relaxant, xylazine, reduces the intensity of the seizure activity and instantly reduces mortality rates. The other critical point in the process is arrest of the seizures. A massive dose of diazepam is currently used, which can stop the heart. Instead, a cocktail of very low doses of synergistic drugs, acting at different receptor systems, can more controllably terminate the seizures, Woodhall explained. It turns out, he said, that acute status epilepticus can be avoided. To make the most of the model, Woodhall identified several ways to increase the use of the fragile brain slices obtained from animals. Enlisting multiple researchers on one day to extract as much data as possible from each individual rat reduces the number of animals needed during the experiment. Methods for production and storage of slices were also improved. Together, these refinements led to development of a new model, low-dose lithium-pilocarpine-xylazine epileptogenesis, with a very brief period of acute status epilepticus, much longer quiescent period, and less than 2 percent mortality. The new model, which mimics the unique features of pediatric epilepsy, was validated using brain slices from children who had surgery for intractable epilepsy, Woodhall noted. In addition to refining the models themselves, Woodhall said that data sharing among researchers is another aspect of refinement and overall reduction as well.

Refinement presents some challenges, Woodhall noted. The new epilepsy model, for example, takes longer to achieve recurrent seizures and is therefore more expensive, and there is more variability. Woodhall concurred with Barnett that it can be challenging to publish refinements to methods that have been broadly used for decades.

SUPPORTING THE 3Rs WITH PRECLINICAL SYSTEMATIC REVIEWS

Clinical systematic reviews combine the results of many different studies, increasing the power of analysis and confidence in the conclusions. Meta-analysis of clinical trials has long been used in drug development to gain a fuller picture of the potential efficacy of an investigational compound. Meta-analysis has, for example, identified shortcomings of individual trials, identified toxicities that were not significant in a single study, influenced how future trials should be designed, and clarified responses of different subpopulations of patients.

Anne Murphy, associate professor at the University of California, San Diego, suggested that systematic reviews of preclinical data and translational animal studies could assist with replacement, refinement, and reduction of animal use in neuroscience research. A systematic review is a formulaic, statistically based approach to analyzing preclinical data. The formulaic approach to systematic reviews minimizes bias and maximizes transparency; the results are objective and quantitative. In general, the steps of a systematic review are

- Conduct exhaustive search for published and unpublished relevant data.
- Select studies for inclusion that meet predetermined criteria.
- Critically appraise studies, evaluate quality, and extract data.
- Combine data and apply appropriate statistical analysis.
- Draw conclusions and write manuscript.
- Update review as additional relevant studies emerge.

Can Systematic Reviews Assist with the 3Rs?

Systematic reviews could also assist with replacement, refinement, and reduction, Murphy suggested. Preclinical systematic reviews could potentially:

- *Replace* animal use by
  - o providing evidence of the validity of studies by comparing in vitro, invertebrate, or in silico data with data from traditional animal studies.
- *Refine* experimental procedures by
  - o highlighting how differing methodologies affect measures of efficacy.
  - o providing a platform for setting a standard for the methodology of a particular model and unifying the reporting requirements.
  - o providing evidence of the effectiveness of refinements.
- *Reduce* the ineffective use of animals by
  - o avoiding duplication, preventing further studies of ineffective interventions.
  - o providing a more precise estimate of treatment effect, thereby informing future power analysis.

Systematic reviews do have weaknesses, however, noted Murphy. The value of the review for the development of therapeutics for humans depends on the quality of included preclinical studies. Systematic reviews can become outdated rather quickly and must be regularly updated as new data become available. This requires some sort of repository or electronic warehouse for the data so that modifications can readily be made. Systematic reviews can still be susceptible to bias in the selection of studies, especially if the predefined rules are not followed. Finally, there is the challenge of obtaining unpublished data; in particular, negative data are difficult to collect.

Preclinical Studies

A fundamental problem with the use of animals in research is that efficacy in animal models of disease does not necessarily equal efficacy in humans, Murphy noted. Many compounds come through animal studies only to fail in the clinic. Clinical trials fail for a variety of reasons. For example, they may be underpowered or they may underestimate the variability of the endpoint measures, leading to inconclusive results. However, sometimes the treatment regimen for humans differs from that of the animal model. In stroke, for example, the majority of preclinical data suggested a short therapeutic window; however, it generally is not possible to see a patient within 15 minutes after the start of a stroke. This, Murphy suggested, is one of the reasons that many stroke compounds have failed in the clinic.

Some animal studies also have methodological bias. As an example of empirical bias in the design of experimental stroke studies, Murphy noted that studies are generally done in young, healthy, male animals, while humans who have strokes are generally older with comorbidities (Crossley et al., 2008).

The quality of preclinical studies is highly variable. A recent survey found that 40 percent of 271 randomly chosen articles did not state a hypothesis or objective, or the number and characteristics of animals (e.g., species, strain, sex, age, weight). The survey also found that more than 85 percent of studies did not report randomization or blinding and 30 percent did not report statistical methods (Kilkenny et al., 2009). Study quality influences measures of efficacy. The assessment of study quality is an inherent part of a systematic review, Murphy noted.

Murphy suggested that preclinical systematic reviews could help address some of these issues. A systematic review by Perel and colleagues (2007) comparing treatment effects in animal experiments and clinical trials found systematic reviews of preclinical data could identify low-quality animal studies and better predict success or failure of compounds in clinical trials.

Integrating Systematic Review into Preclinical Translational Research

In summary, Murphy said, systematic review could be applied to preclinical data in order to improve the overall quality and value of animal studies, support the 3Rs, and inform clinical trials. The path to implementation of systematic reviews as a matter of routine potentially includes the Food and Drug Administration, pharmaceutical companies, research institutions, and publishers.

Murphy suggested two strategies for supporting systematic reviews of preclinical research. The first is to raise awareness of the power of applying systematic reviews to animal studies. Conducting reviews can inform and improve the timing, design, and quality of studies and better inform subsequent clinical trials. The second is to secure support from publishers and journal editors. Access to useful data might increase with more rigorous application of requirements for publication and rejection of low-quality or incomplete studies. In addition, the support for the publication of negative data would enable increased sharing of primary data, regardless of the outcome.

IMPACT OF THE 3Rs ON DRUG DISCOVERY AND DEVELOPMENT

Jackie Hunter of OI Pharma Partners discussed how the evolving pharmaceutical industry may change animal research, specifically, how human studies could lead to opportunities for increased application of the 3Rs and how changes in business models could lead to greater data sharing and hence, opportunities for reduction in the numbers of animals used as well. The pharmaceutical industry faces many challenges in bringing a new product to market. Hunter noted that over the past 10 to 15 years, the number of approvals of new drugs for nervous system disorders has dropped and the pharmaceutical industry in general is moving away from neuroscience research.

Refinement Stemming from Target Validation in Humans

In drug development, animal research plays a role in target validation, screening of compounds to optimize pharmacokinetics and efficacy, and safety and toxicology testing. Advances in technologies, however, are enabling increased target validation in man, potentially reducing the need for animals. Studies of the genetics of rare diseases, imaging studies, genome-wide association studies (GWAS), pharmacogenomics, and stem cell research are informing industry decisions to pursue particular drug targets.

For example, researchers are modeling schizophrenia using human-induced pluripotent stem cells, identifying new pathways and potential drug targets that have not been previously associated with schizophrenia (Brennand et al., 2011). Studies of mutations in individuals with rare diseases or isolated syndromes who exhibit a gain or loss of function also can help focus drug discovery efforts. For complex disorders involving multiple genes, GWAS are beginning to cluster pathways, identifying convergent nodes on these pathways that may be important in terms of disease progression.

While increased target validation in humans is unlikely to replace animal models, it will allow refinement of the questions asked of the models, Hunter said. For example, knowledge from human validation studies could lead to an increased focus on models of mechanism, rather than models of disease. Refined models could also help identify unwanted target-related effects, allowing a target to be invalidated early in the process.

Animal Models of Mechanism Versus Models of Disease

Few animal models faithfully represent the full complexity of the disease being modeled. This is especially true for nervous system disorders and diseases, for which animal models are limited in predicting drug efficacy, Hunter noted. Animal models may provide conflicting data in terms of exposures of drugs required, and result in false negatives. As a result, products are frequently tested in multiple models.

Hunter suggested a need to move toward more mechanistic models. Increasing disease knowledge allows for better identification of key mechanisms. The focus then should be on developing mechanistic in vivo assays that can be translated to humans. Such assays could demonstrate compound effects on the mechanism, define the exposures required for efficacy on the mechanism, and allow comparison of pharmacodynamics with pharmacokinetics. This could lead to a reduction in the number of models and experiments needed, Hunter opined.

This approach requires a different mindset, Hunter said. For progression into human trials, if a molecule works in an animal model of disease, it is often necessary to show that it works in several models. On the other hand, if a molecule works on a particular mechanism, only one experiment may be needed to take it forward.

Precompetitive Collaborations

The current economics of drug development are not sustainable, Hunter commented. One approach to help move discovery forward is the concept of precompetitive collaborations. A number of efforts are underway globally to share more data and information. One example is the Innovative Medicines Initiative (IMI),[1] a public–private partnership between the European Federation of Pharmaceutical Industries and Associations and the European Union. Large consortiums facilitated by IMI share information on existing animal data, developing new models, and standardizing models across different companies and institutions. The NEWMEDS Consortium, for example, is working to develop both new preclinical models and translational experimental medicinal models for schizophrenia and depression.

In summary, Hunter stressed that advances in technology and creative approaches to precompetitive collaboration and data sharing are providing real opportunities to refine animal models.

---

[1] See http://www.imi.europa.eu/.

BOX 5-2
Summary of Session Points

3Rs
- Advances in technology have and will continue to provide opportunities for replacement, refinement, and reduction (3Rs).
- Increased understanding of disease mechanism may help in development of replacement and refinement strategies.
- *Replacement*: In vitro cell culture models can be used to test reagents and potential therapeutic candidates, including prescreening combinations of biological and pharmacological agents.
- *Refinement*: Simple refinements can improve study results while positively impacting concerns about animal care and use.
- *Reduction* is often a "by-product" of refinement.

Systematic Reviews
- Systematic reviews of preclinical data could potentially:
  - Improve the quality and value of animal studies and support the 3Rs.
  - Better inform the timing, design, and benefit of clinical trials.
- The path to implementation of systematic reviews of preclinical data might include the Food and Drug Administration, pharmaceutical companies, research institutions, and publishers.

SOURCE: Individual panelists and participants.

6

Public Engagement and
Animal Research Regulations

Animal research regulations, laws, and policies undergo continual review and revision as new technologies, tools, and techniques alter the way in which animals are used in biomedical research. The process of updating regulations most recently occurred with the revision of the European Union (EU) directive. As this process demonstrated, the development and implementation of new or modified regulations are also directly impacted by changes in public opinion. Given this, any discussion of the reasons for the establishment of animal research regulations, their impact on research, and opportunities for harmonization includes an examination of how the use and regulation of animals in neuroscience research is communicated with the public. As MacArthur Clark noted, public opinion is an important component of a balanced regulatory system that takes into account public confidence, scientific quality, and assurances of animal welfare.

With this in mind, the session, which included panelists from academia, patient advocacy, and the media, discussed engagement of the public, politicians, and the media on animal research issues, and opportunities to communicate with non-researchers on the regulation of the use of animals in neuroscience research. (Key points are summarized at the end of the chapter in Box 6-1.)

REACHING OUT FROM ACADEMIA

Randall Nelson, professor and associate vice chancellor for research at the University of Tennessee Health Science Center, shared his perspective as a mid-level administrator. He described what is a bit of reluctance on the part of investigators to engage the public and the press about both animal use in research and the regulations that govern the use. Some fear that discussing their research will make them targets and they perceive the potential for harm as very real, noted Nelson. In addition, the benefits of engaging the public in these issues may not always be immediately tangible to scientists. Nelson informally surveyed his colleagues and found that many perceive the scientific press to be understanding, rational, accurate, and unbiased. Others suggested that the general press, however, was negatively perceived by some, aligned with specific interests, and interested in the story, but not necessarily in accuracy. Scientists are required to understand the regulations and policies governing animals in research. Many of the scientists who spoke with Nelson believe the press does not fully understand these regulations, but that they have a responsibility to "get it right" when reporting to the public. The question then, Nelson asked, what is "right"?

Scientific progress depends on public perception and acceptance, Nelson said, and he offered several observations and suggestions regarding methods to increase opportunities for success in engaging the public:

- **Public confidence may increase with public dialogue.** Information fosters understanding, and understanding fosters appreciation. The cycle of mistrust must be interrupted for progress to be made, and who better to do that than those directly involved in the science, Nelson asked.
- **Increasing opportunities for interfacing with the public.** Frank discussions with the public about animal research regulations and the use of animals in science may dispel the notion of "ivory towers and dark secrets which reside within research laboratories."
- **Train scientists to communicate clearly about animal research.** Having someone else communicate for the scientific community dilutes the message, Nelson noted. Connecting a scientist's name and face with a news item could help foster a relationship and build trust. In speaking about their research, Nelson suggested that scientists should
  o Give general examples rather than elaborating on specifics.
  o Consider the audience's previous knowledge when delivering information.
  o Describe efforts by scientists, regulators, and government agencies to minimize animal numbers and pain and distress.
  o Help people understand the process of cost-benefit analyses when using animals in research.
  o Ensure accurate representation of the science. Public confidence may increase if the public has access to accurate information.

   o   Explain why an animal model was chosen, and the drawbacks
       of other approaches. While scientists often discuss why a par-
       ticular model was chosen, they rarely talk about why other
       possibilities were rejected.
   •   **Engage patients and patient advocacy groups.** People living with
       disease hold particular interest in learning about new research.
   •   **Promote appropriateness of use of animal models within the scientific
       community.** Regulations require justification of the use of animals.
       However, scientists assume a moral and ethical responsibility when
       agreeing to do animal research, Nelson said, and should understand
       that the use of animals in research is a privilege, not a right.

MEDIA COVERAGE OF ANIMAL RESEARCH

   Mark Henderson, science editor for *The Times* in London, described
how the media discourse on animal experimentation has changed in the
United Kingdom. Ten to 15 years ago, there was a significant amount of
opposition to the use of animals in research, he said, and few scientists who
engaged in animal experimentation were willing to publicly discuss their
research. This meant journalists had few sources for a scientific viewpoint
on studies involving animals.
   Henderson noted that media coverage at the time tended to report
negative representations of animal research more often. These included the
use of graphic images of animals during experimentation. Henderson noted
that many of the pictures published were taken quite a long time ago and
from sources outside the United Kingdom.
   Some of the media, he noted, has always tried to portray animal research accurately, but this has proven difficult as again, there were few
scientists willing to discuss the use of animals. It was also difficult to obtain
up-to-date images of the work that was actually done in labs. As a result,
the political climate toward animal experimentation in the United Kingdom
was lukewarm, Henderson noted.
   Around 2002 to 2003, a number of things came together that changed
the conversation about animal research in the United Kingdom. The Science Media Center was formed to support scientists working in engaging
the public.[1] From a "safety in numbers" perspective, a single voice became the immediate magnet for the protests, but hundreds of researchers
talking about animal research made it much harder to single individuals
out, Henderson observed.
   Nelson mentioned that individuals living with disease are knowledge-
able about the use of animals in research and can also take part in public

---

[1] See http://www.sciencemediacentre.org/pages/.

engagement efforts. Henderson cited the contribution of the late Laura Cowell, a cystic fibrosis patient who as a teenager in 2002 began speaking about her support for animal research because it was essentially keeping her alive. Cowell's efforts had a strong effect because it personalized the issue, making the animal research story one that was about actually helping people. A significant secondary effect, Henderson noted, was that it also emboldened scientists.

A new trend to discuss the use of animals in research, both with the media and directly with politicians, started to develop among scientists. Around 2005, Henderson said he began to receive invitations to visit laboratories conducting research with animals, so he could see firsthand what was involved. One recent development as a result of the increased dialogue between scientists and the media is that the use of animals in research is now described incidentally in the high-end British media (e.g., *The Times*, *The Guardian*, and the BBC), Henderson noted. For example, recent coverage in *The Guardian* of a study published in *Nature* on stem cells and cardiac repair, casually mentions in the article subtitle the fact that mice were involved in the research. There is no controversy; it is just presented as a fact of the study.

Henderson noted that more could be done to increase dialogue concerning the use of animals in science. For scientists who do research involving animals, including information about the role of the animals when they talk to the media could be one method. Do this, he stressed, in an incidental, normative fashion. The more that researchers discuss animals and animal regulations as an integral part of medical research, the more people will actually become aware. Engagement works, Henderson said. Public opinion over the 10 years in which the scientists have made a real effort to engage has increased steadily. He went on to say that unconditional support of the statement "research on animals is acceptable" has doubled in 10 years, from 32 percent in 1999 to 60 percent in 2009, and conditional acceptance hovers around 85 to 90 percent in the United Kingdom.

The relationship between science and the media is not symmetrical, Henderson observed. Scientists have a responsibility, particularly if they are publicly or charitably funded, to talk publicly about their research. Journalists do not have a responsibility or duty to be accurate (although many strive to be). Journalists want to tell stories and biomedical research has an excellent story to tell. Henderson suggested that scientists talk about animal research accessibly and often, without exaggerating the benefits.

THE ROLE OF PATIENTS AND PATIENT GROUPS

Timothy Coetzee, chief research officer of the National Multiple Sclerosis (MS) Society, discussed animal research regulations from the patient advocate perspective. The National MS Society invests about $40 million in research, funding about 325 projects worldwide, at least a third of which have an animal approval associated with them.

Animal models will continue to be necessary in MS research for the foreseeable future, Coetzee noted. Due to the nature of the disease, alternatives to some of the models for drug discovery and development are not currently possible. Coetzee described some of the animal models used in MS Society–funded research. The experimental allergic encephalomyelitis (EAE) model in a non-human primate is as close as science has come thus far to an animal model for MS. Few labs use it because of its difficulty and Institutional Animal Care and Use Committee (IACUC) concerns about the model's profoundly debilitating effects. There are viral demyelinating models (Theiler's virus, mouse hepatitis virus), that mimic progressive forms of the disease. The problem, particularly with the Theiler's model, is that few pharmaceutical companies want to work with a viral demyelinating model; it is not clean and no one wants to introduce viruses into their animal colony. Other researchers have used chemically induced demyelination (e.g., cuprizone, lysolecithin, ethidium bromide). This is not a model of MS, as there is no autoimmune inflammation, only demyelination. Regardless, these models are powerful tools for understanding how to rebuild myelin in the nervous system. Coetzee noted that the MS patient community supports the society's investment in animal research because they believe that without it, current therapies would not be available.

The National MS Society's policies on animal research are relatively straightforward, Coetzee explained. Applicants for funding are required to provide the Society with an IACUC approval (or the equivalent outside the United States) before any funds are released. The Society also conducts annual monitoring.

Overall, the National MS Society's philosophy is that animal research is essential for progress in MS research. The society encourages the development of alternative models. The society also publicly states its support for animal research, but is careful not to draw attention to animal research unless needed.

Session chair Frankie Trull opined that defending and supporting animal research is not necessarily the same as promoting the best outcomes for human health, and suggested that scientific scrutiny of animal experiments could be a role for patient advocacy groups. Coetzee responded that the National MS Society does consider whether an animal model is appropriate when granting funds and agreed that this is an element of oversight that patient advocacy groups can apply. Patient advocate organizations, as well as commercial organizations, government, and all funders, bear a responsibility to ask what the costs of these cures and treatments developed.

As patient advocates, Coetzee said, patient groups have to discuss animal research. As an organization that funds research, the National MS Society has a responsibility for effectively communicating this research to the community. The public can appreciate the nuances of animal research, he said, but communication is most effective when it is very brief and very focused. Coetzee suggested that scientists think in terms of headlines when preparing to discuss research. It is also important to pay attention to shifting approaches in media, especially social media.

BOX 6-1
Summary of Session Points

Scientists and Institutions
- Investigators may be reluctant to engage the public and the media. Many investigators:
  - fear that discussing their work involving animals will make them targets;
  - do not see the benefit of engaging the public; and/or
  - see the general press as biased.
- Institutions could invest in training and equipping scientists to speak to the media and the public.

The Media
- Building individual relationships with journalists and media might increase communication between the groups.
- Education is not necessarily a responsibility of mainstream media.
- Some media outlets seek to attract customers by earning a reputation for being trustworthy and accurate.
- More could be done to increase dialogue about the use of animals in science.

Patient Groups
- People living with disease can be the best advocates for disease prevention and cure.
- Patient advocacy groups could serve an oversight role in scrutinizing the scientific value of projects using animals before choosing to fund them.

Engaging the Public
- Public engagement and education can increase support for use of animals in research.
- Communication may be most effective when it is brief and focused, with the role of animals mentioned incidentally.

SOURCE: Individual panelists and participants.

7

Core Principles for the
Care and Use of Animals in Research

In this session, government, industry, and other animal research regulation stakeholders discussed the feasibility of developing core principles governing the use of animals in research (key points and overlapping core principles are summarized at the end of the chapter in Box 7-3). To start the discussion, Richard Nakamura, session chair and then scientific director of the National Institute of Mental Health, presented his personal perspective on the use of animals and animal welfare.

BALANCING SCIENTIFIC PROGRESS AND ANIMAL WELFARE

The goal of biomedical research is to understand living systems, with a particular focus on human biology and human disorders, Nakamura said. Scientists can learn much about the principles of human biology and behavior from animal models and use discoveries in human biology to help understand other animals. Animals are studied instead of humans in many situations where manipulating humans is not possible or acceptable. The use and sacrifice of these animals bred for research might be considered an acceptable ethical cost *if* there is appropriate consideration for their welfare in life. A key welfare consideration for animals in research, Nakamura said, is to minimize pain and distress and improve overall well-being.

The ethical challenge to conducting animal research is balancing the gains in scientific knowledge with the costs to animals, especially in terms of pain and distress. While there is no satisfactory calculus for doing this, scientists and others have tried to reach an acceptable balance by using approaches such as the 3Rs (replacement, refinement, and reduction),

Nakamura said. This, however, may have the counterproductive effect of limiting scientific gains from studies while increasing the costs. Because many medical needs remain unmet and much needs to be learned about living systems and disorders, Nakamura said the growth of welfare considerations and regulations must be constrained to allow some proportionality to the world outside of research (e.g., animals used for food, organisms displaced by humans or killed as pests). Nakamura suggested that animal welfare should not be considered in isolation from scientific goals or the larger needs of society.

The 3Rs are used when considering approval of animal research studies. However, the 3Rs are not a core principle, because as a core principle the 3Rs, specifically the principle of replacement, would translate into a goal to end animal research, Nakamura observed. If people are not ready to apply this to the world outside research, including the eating of meat or killing of pests, it should not be a core principle in research, Nakamura opined.

The key principle of animal regulation in research might be finding a balance among scientific progress, animal welfare, and cost effectiveness that is better than, yet proportionate to, the larger treatment of animals by humans, Nakamura concluded.

EUROPEAN REGULATORY PERSPECTIVE

Judy MacArthur Clark of the UK Home Office discussed core principles and considerations from a European regulatory perspective. She reiterated that a goal of regulatory balance might be to provide the public with confidence that animals will be appropriately protected and that science will not be inhibited from discovering solutions to many global health problems (see Chapter 2). While public opinion polls on animal research informed the development of the directive, MacArthur Clark noted that there is little direct support in the directive of high-quality science in Europe.

Core Principles

According to MacArthur Clark, three basic core systems are necessary in animal research regulations: a system of authorization of people, places, and projects; ethical, impartial, and independent evaluation of projects based on a cost-benefit balance; and impartial and independent verification of compliance involving some form of inspection.

MacArthur Clark referred participants to Chapter 7.8 of the World Organisation for Animal Health (OIE) *Terrestrial Animal Health Code*, which provides basic core principles of a regulatory system framework for use of animals (OIE, 2011). Per this code, components of the systems described above would include implementation of the 3Rs, evidence of training and competence of individuals (ethical, legal, and specific skills), provision of veterinary care, viable sources of animals and transport, and inspection of facilities.

There are consequences of applying core principles, MacArthur Clark suggested. First, some proposals may not be considered justifiable. They may, for example, involve long-lasting, severe pain that cannot be alleviated or the cost-benefit assessment may not balance. Second, bureaucracy may become part of the evaluation of "marginal" projects. When challenged with a difficult decision, assessors ask more and more questions, resulting in what is sometimes called "paralysis of analysis." Eventually, either the project is approved or the applicant gives up. If the project is approved, it is often on the basis that the work will be heavily monitored, MacArthur Clark noted. This process occurs equally in different models of governance (both the U.S. self-regulated Institutional Animal Care and Use Committee [IACUC] model and the European Union [EU] centralized Competent Authority).

Project Authorization Without Bureaucracy

MacArthur Clark raised the concept of "thin slicing" as an approach to making project authorization decisions. The concept is that spontaneous decisions are often as good as, or even better than, carefully planned and considered ones. The goal would be to make better decisions early on and foster compliance without bureaucracy. Another goal would be to identify marginal projects as early as possible and develop a separate process to review them and facilitate faster decisions. Reject the project and provide reasons, or accept the marginal project and build in milestones that are not overly restrictive, MacArthur Clark asserted. Once the decision to authorize has been made, it might be easier to focus on other aspects of approval, such as implementing the 3Rs. MacArthur Clark noted that efficient processing could reduce costs while supporting science and welfare.

Training and Competence

From the outset, the European commission declared its intention that there should be free movement of staff, including scientists, veterinarians, and animal care staff, throughout Europe. This means a need for common training standards. Training, however, does not necessarily equate with competence, MacArthur Clark noted. Individuals acquire competence through their work, so in addition to training and supervision, a mechanism to maintain competency could be included.

Another core principle could be the delivery of appropriate and relevant training that is acceptable and palatable to those who receive it.

Such a core training program could cover legislation, ethics, and the 3Rs. Recall that each EU member state has its own legislation, so even though staff can move freely among countries, they might need to understand the local legislation and how it is implemented. There could also be modules of additional training appropriate to the particular needs of the individual and continued training to maintain developing skills.

Harmonization and Consistency

Harmonization is a key aim of the EU directive, MacArthur Clark noted. Portions of the directive involve significant changes for some of the 27 EU member states. Areas of harmonization across the member states might include the types of animals that are protected; minimum housing standards; project authorization and severity classification; accountability for training and competence requirements; animal welfare body and a designated veterinarian; and some form of inspectorate for compliance. Member states will be allowed to apply higher standards, as long as they do not interfere with market competitiveness. Consistency is very important, and could be considered a core principle, MacArthur Clark noted. Harmonization is a step in the right direction, but is not likely to achieve total consistency, in large part because each member state has its own legal system. The ability for member states to apply different standards may impact harmonization of regulations. The extent of harmonization might not be known until all member states transpose the directive in 2013.

In closing, MacArthur Clark quoted from the directive:

> This Directive represents an important step toward achieving the final goal of full replacement of procedures on live animals for scientific and educational purposes as soon as it is scientifically possible to do so. (European Union, 2010)

This "full replacement" is a goal of the directive and would lead to the discontinuation of animal research. MacArthur Clark stressed that it is important that scientists in Europe continue to speak up about the importance of animals in biomedical research.

U.S. REGULATORY PERSPECTIVE

Patricia Brown, director of the Office of Laboratory Animal Welfare at the National Institutes of Health (NIH), discussed U.S. government principles as they apply to research funded by the NIH. The U.S. system is different in that NIH-supported research is a partnership with the grantee institutions. Brown noted that both sides share a mutual need for compli-

CORE PRINCIPLES FOR THE CARE AND USE OF ANIMALS IN RESEARCH   65

ance with the laws, regulations, and policies that apply to animal research. Each partner has responsibilities and obligations as stewards of public funds. Institutional self-governance is the foundation of the Public Health Service (PHS) Policy, and all research institutions receiving funds within the United States must agree to follow the U.S. Government Principles for the Utilization and Care of Vertebrate Animals Used in Testing, Research, and Training.[1]

U.S. Government Principles for the Utilization and Care of Vertebrate Animals Used in Testing, Research, and Training

Brown reviewed the aspects of the three primary entities involved in oversight of animal use in the United States.[2] The PHS Policy requires an institutional program of animal care and use, with an IACUC appointed by the chief executive officer. There must be an institutional official who is responsible for the general administration of the program, including the provision of resources to support the animal care and use program. The IACUC and the attending veterinarian must report to an institutional official. Institutional self-regulation is based on the PHS Policy, the *Guide for the Care and Use of Laboratory Animals*, and if there are regulated species in the program, the Animal Welfare Act regulations.

The U.S. Government Principles for the Utilization and Care of Vertebrate Animals Used in Testing, Research, and Training were formulated in 1985, and formed the foundation for the Animal Welfare Act Amendments and the PHS Policy. They are basic tenets of animal care and use, Brown said. The expectation of the PHS Policy is that IACUCs use the U.S. Government Principles as the basis for protocol review. Brown highlighted several principles that illustrate where the United States is in terms of animal care and use (Box 7-1).

International Guiding Principles for Biomedical Research Involving Animals

The International Guiding Principles for Biomedical Research Involving Animals were issued in 1985[3] and were the basis for the development of the U.S. Government Principles. They were developed by the Council of International Organizations of Medical Sciences (CIOMS) and are endorsed by both the European Medical Research Council (EMRC) and the World Health Organization (WHO).

---

[1] See http://grants.nih.gov/grants/olaw/references/phspol.htm#USGovPrinciples.

[2] See discussion by Bennett in Chapter 2 for further detail on the PHS Policy and the Animal Welfare Act.

[3] See http://www.cioms.ch/publications/guidelines/1985_texts_of_guidelines.htm.

BOX 7-1
Excerpts from the U.S. Principles

Principle II Rationale: "Procedures involving animals should be designed and performed with due consideration of their relevance to human or animal health, the advancement of knowledge, or the good of society."

Principle III Justification: "The animals selected for a procedure should be of an appropriate species and quality and the minimum number required to obtain valid results. Methods such as mathematical models, computer simulation, and in vitro biological systems should be considered."

Principle IV Minimize Pain and Distress: "Proper use of animals, including the avoidance or minimization of discomfort, distress, and pain when consistent with sound scientific practices, is imperative. Unless the contrary is established, investigators should consider that procedures that cause pain or distress in human beings may cause pain or distress in other animals."

Principle VI Humane Endpoints: "Animals that would otherwise suffer severe or chronic pain or distress that cannot be relieved should be painlessly killed at the end of the procedure or, if appropriate, during the procedure."

Principle VIII Training: "Investigators and other personnel shall be appropriately qualified and experienced for conducting procedures on living animals. Adequate arrangements shall be made for their in-service training, including the proper and humane care and use of laboratory animals."

Principle IX Ethical Review: "Where exceptions are required in relation to the provisions of these Principles, the decisions should not rest with the investigators directly concerned but should be made . . . by an appropriate review group such as an IACUC. Such exceptions should not be made solely for the purpose of teaching or demonstration."

SOURCES: Brown presentation, U.S. Government Principles for the Utilization and Care of Vertebrate Animals Used in Testing, Research, and Training.

More recently, the International Council for Laboratory Animal Science (ICLAS) and CIOMS convened a working group to revise and update the International Guiding Principles. A draft was released for comment in April 2011 and the final draft is expected in 2012.[4] Among the proposed revisions, the draft now states that the use of animals is a privilege that carries with it a moral obligation. Brown quoted from the draft revision

---

[4] See http://ora.msu.edu/ICLAS/topics.html.

to Principle II: "Individuals working with animals have an obligation to demonstrate respect for animals, to be responsible and accountable for their decisions and actions pertaining to animal welfare, care, and use, and to ensure that the highest standards of scientific integrity prevail." The 3Rs are now specifically mentioned in Principle III whereas previously they were only implied. Principle VII more fully addresses pain and distress in animals and calls for consultation with a veterinarian. Principle VIII calls for establishing endpoints before animal use begins, with assessment throughout the course of the study. Finally, Brown noted that the idea of risk-benefit analysis for animal use, balancing the benefits derived from the research with the potential for pain and distress experienced by an animal, is now addressed in the draft Principle X.

PHARMACEUTICAL PERSPECTIVE

Margaret Landi, vice president of Global Laboratory Animal Science and chief of Animal Welfare and Veterinary Medicine for GlaxoSmithKline (GSK) Pharmaceuticals,[5] noted that a challenge for multinational corporations is developing an animal care system that meets the required standards, but that also recognizes the global diversity of rules, regulations, policies, and guidance. She also pointed out that while pharmaceutical companies have extensive internal capabilities, they also have numerous external collaborations, contracts, and alliances.

Landi noted that developing global guidance on animal care and use can be hampered by a lack of consensus on what "best practice" is. Lack of regulatory harmony, evolving regulations and guidelines, and cultural differences can add to the challenge.

Landi questioned whether variations in rules, regulations, policies, and practices really result in a substantive difference in animal care, and suggested that alignment of principles can be achieved independent of differing practices.

GSK, Landi said, is aligned to follow the guiding principles:

- **Performance standards:** Identify clear accountability, responsibility, and desired outcome. Standards describe the desired outcome, but allow flexibility in achieving the outcome.
- **Professional judgment:** The ability of a person to recommend a course of action due to specialized knowledge and/or skills.
- **Harmonized (consistent) approaches or outcomes:** For example, one consistent approach is a program of veterinary care; however,

---

[5] GSK is headquartered in the United Kingdom, with facilities in North America, Spain, France, China, and Singapore.

what that program looks like may differ at each company site because of the availability of drugs in different countries, or their use
of different models or species at different sites. Standard operating
procedures foster consistency.

Ultimately, the goal would be to articulate a set of core principles for
care of animals within and for the institution. Landi stressed that difference in practices do not necessarily equate to difference in care or "dual
standards."

As noted above, multinational companies can influence animal care and
use practices in regions where they conduct research. Landi shared the GSK
core principles for the use of animals in research (Box 7-2), which apply to
both in-house research and external collaborations, regardless of location.

DEVELOPING CORE PRINCIPLES

Timo Nevalainen, from the National Research Council's Institute of
Laboratory Animal Research (ILAR) Council, said there are several potential target populations for core principles of animal research: specific fields
of research (e.g., neuroscience), institutions, regulators, and the general

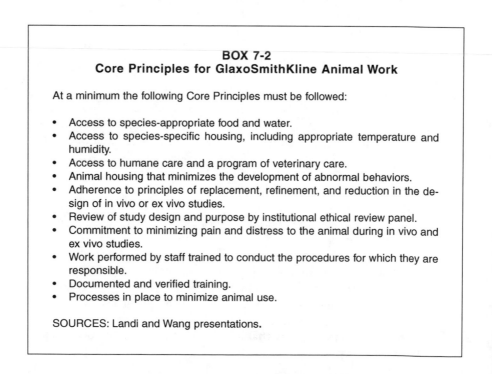

BOX 7-2
Core Principles for GlaxoSmithKline Animal Work

At a minimum the following Core Principles must be followed:

- Access to species-appropriate food and water.
- Access to species-specific housing, including appropriate temperature and humidity.
- Access to humane care and a program of veterinary care.
- Animal housing that minimizes the development of abnormal behaviors.
- Adherence to principles of replacement, refinement, and reduction in the design of in vivo or ex vivo studies.
- Review of study design and purpose by institutional ethical review panel.
- Commitment to minimizing pain and distress to the animal during in vivo and ex vivo studies.
- Work performed by staff trained to conduct the procedures for which they are responsible.
- Documented and verified training.
- Processes in place to minimize animal use.

SOURCES: Landi and Wang presentations.

public. Nevalainen raised several questions for consideration. First, should the scientific community be proactive and establish core principles for the regulators? One example of core principles of animal welfare developed by scientists and referred to by government policy makers is the ILAR *Guide for the Care and Use of Laboratory Animals* (NRC, 2010).

Next, how specific should core principles or regulations be? The EU animal care and housing regulations, for example, are species specific, but do not go to the level of stocks and strains. Strains differ in many ways, such as how they react to environmental enrichment items placed in the cage. Optimally, Nevalainen said, housing and care regulations or guidelines could be specific to the level of strains and stocks, but this may be unrealistic due to the vast number of different strains. In the EU directive, there is some division by type of research (e.g., basic, translational research, regulatory testing for quality, efficacy, and safety), but not by field.

Third, would regulations specific to neurosciences be helpful? Nevalainen noted that in 2003, ILAR issued *Guidelines for the Care and Use of Mammals in Neuroscience and Behavioral Research* (NRC, 2003).

Focusing on Refinement and Reduction

Nevalainen discussed a "2Rs" approach, seeking to maximize both refinement and reduction, but not replacement, to achieve less harm and better science. A 2Rs method could be scientifically validated, beneficial to the animals, and not detract from the scientific integrity. Even small changes to improve conditions for individual animals might have considerable impact overall, Nevalainen noted. For example, habituating animals to handling can reduce their anxiety and stress.

CORE PRINCIPLES FOR ANIMAL USE IN NEUROSCIENCE RESEARCH

During the discussion that followed the presentations, several panelists and participants observed that the core principles described above are generally applicable to all research areas and procedures involving animals. Specific recommendations or guidance on neuroscience procedures may be needed, but the core principles by which animal studies in neuroscience can be conducted might be the same as those for any discipline. Citing the GSK example discussed above, few participants noted that the basic principles of performance standards, professional judgment, and a harmonized approach or outcome is independent of the therapeutic area being investigated or the pathway being explored. It also was noted that with so many disciplines and research areas, it is unlikely that a directive or legislation could specifically address individual research areas (e.g., neuroscience).

BOX 7-3
Summary of Session Points

- Target populations for core principles of animal research may be specific fields of research, institutions, and regulators.
- Considerations for animals in research could include
  - Welfare: Minimize pain and distress and improve overall well-being.
  - Regulatory balance: Deliver public confidence that
    - animals are appropriately protected.
    - science will not be inhibited from discovering much-needed solutions to global problems.
- Overlapping core principles from presentations:
  - Humane care: Minimizing pain and distress and improving overall well-being.
  - Unbiased ethical evaluation of projects (balancing the cost-benefit ratio).
  - Training and maintenance of a competent workforce.
  - Adherence to principles of replacement, reduction, and refinement.
- Alignment of principles could be achieved independent of differing practices. Difference in practice does not necessarily equate to difference in care or "dual standards."
- Although recommendations or guidance on neuroscience procedures may be needed, the core principles by which animal studies in neuroscience could be conducted might be the same as those for any discipline.

SOURCE: Individual panelists and participants.

8

Summary of Workshop Topics

The topics highlighted in this chapter are based on closing remarks made by each of the session chairs during the final session. Additional comments by session participants that related to those closing remarks were included as noted. Comments should not be construed as reflecting on any group consensus or endorsement by the Institute of Medicine.

SESSION I: INTERNATIONAL ANIMAL RESEARCH REGULATIONS

Global Harmonization

Well-established regulatory systems in the United States and European Union may seem disparate at first, but various speakers noted, this is due more to different terminology than to real differences in principles and outcomes, session chair Judy MacArthur Clark observed. For example, it was discussed that the United States has no standardized animal study proposal form comparable to the UK project license form. However, what must be included in protocols submitted for Institutional Animal Care and Use Committee (IACUC) approval is standardized. Institutions have the flexibility to adapt forms and systems to the intricacies of their own animal research program as long as they meet the standards required by regulation.

Emerging systems, such as those in Brazil and across Asia, are learning from both the successes and challenges of established systems. MacArthur Clark suggested that international collaborations and multinational companies are helping to drive regulations and raise standards in emerging regions, and contribute to global harmonization.

Costs

Bureaucracy in regulatory systems is a challenge shared by those involved in animal research; however, it may be more of a hindrance in some countries than others, MacArthur Clark noted. Regulatory systems have costs that include finances and time for regulators and scientists (i.e., administrative burden). Some participants noted the need to reduce bureaucracy that could impact the progress of science.

Several participants noted that opening a dialogue with the public and politicians about the scientific process, the role of animals in research, and the animal research regulations in place would be beneficial. Other participants noted the need to develop appropriate goals/metrics of success of animal welfare measures to ascertain whether increased costs and burdens result in improved animal welfare.

SESSION II: FREEDOM OF INFORMATION LAWS

Session chair Arthur Sussman observed that there seems to be strong support among workshop panelists and participants for a regulatory environment that is both ethical and intelligent. In addition to regulations specifically governing the use of animals in research, various participants noted that other laws impact research (e.g., animal rights laws, the Freedom of Information Act [FOIA], state sunshine laws, and the False Claims Act). Several presentations discussed how individuals and groups are using FOIA and sunshine laws to obtain information on principal investigators, grants, and matters of noncompliance. In some cases, session speaker Richard Cupp noted that the courts have found connections between release of information under FOIA and subsequent criminal activity against animal researchers. However, it was noted that exemptions to the release of such information are becoming infrequent. Workshop participants noted that increased transparency might not result in decreased information requests, and that transparency needs to be balanced with potential risk. Several panelists and participants emphasized that the suggestion of exercising care when corresponding about animal use in research might be worth particular consideration by scientists.

SESSION III: THE ROLE FOR ANIMALS IN NEUROSCIENCE RESEARCH

Much of the discussion on the impact of animal research regulation on neuroscience research focused on the use of non-human primates, noted session chair Roberto Caminiti. Panelists outlined the current role for non-human primates in biomedical research, including neuroscience research.

It was noted that primate studies can complement in vitro studies, in silico modeling, human brain imaging, and parallel investigations in rodents.

A few participants described how public pressure and politics have phased out certain fields of research in some countries by making it difficult, if not impossible, to study non-human primates. The high cost of using non-human primates was also indicated as creating difficulties in developing this field. Many participants acknowledged, however, that new regulations and laws have produced a significant improvement in animal care, which has led to improved science.

Refinements to animal models used in neuroscience have resulted in reductions in the number of animals required for a study, making the use of non-human primates more feasible. Although long-term data collection from an individual animal allows for the use of fewer animals, it includes numerous independent assessments, which have raised the issue of reuse and severity classification.

New regulations present a host of challenges for institutions, investigators, and IACUCs. Several participants raised concerns about regulations that result in increased costs of conducting biomedical research, without evidence that additional regulations result in improved animal welfare.

SESSION IV: REPLACEMENT, REFINEMENT, AND REDUCTION IN NEUROSCIENCE RESEARCH

Session chair Colin Blakemore briefly reviewed the session's focus on the 3Rs (replacement, refinement, and reduction), the ethical framework for the human use of animals in research. Panelists described an in vitro model of spinal cord injury that replaces the use of animal models and a new model of epileptogenesis that relies on refining previously described techniques. Some participants noted that both of these models have led to the reduction of animal use in experiments. Implementation of the 3Rs principle has a positive impact on improving neuroscience research, noted several participants, as demonstrated by the studies presented.

Publication in the peer-reviewed literature is the primary way information is disseminated in the scientific community. Blakemore observed that a few panelists, however, faced challenges in getting refinements to animal models or replacement methods published, especially when the submitted manuscript proposed to refine or replace an animal model that has been used broadly for decades.

Systematic Reviews

Some participants discussed how systematic reviews of preclinical data could potentially support the 3Rs, thereby improving the quality of animal studies. Participants also noted that systematic reviews provide information about the validity of preclinical data for advancing therapeutics for humans. The value of a systematic review depends on the quality of studies included. Many participants stressed the need for preclinical animal data to be more accessible, including primary data, precompetitive data, and negative data. Establishing repositories of preclinical data is one approach, and issues of funding, location, oversight, and access were raised. In addition, several panelists and participants called on publishers and journal editors to examine current publication requirements for animal studies in hopes of improving the quality of published studies involving animal models.

SESSION V: BUILDING PUBLIC CONFIDENCE

Many participants stressed the importance of transparency, or being honest and open with the public about the regulations, laws, and policies that govern the use of animals in biomedical research. Some participants noted that educating the public about the societal benefits of research using animals is important, but potential harms to the animals is often not discussed. Many participants acknowledge that the public may already understand the value of animal studies and their concerns may be focused more on the quality of life of the animals. Explaining how the care and the husbandry of the animals is managed might be of more interest to the public than scientific details of research studies.

Scientists Talking Science

Session chair Frankie Trull emphasized that scientists have a responsibility to inform and educate the public. Several participants noted that neuroscientists may have an additional responsibility to talk more practically and pragmatically about public outreach programs, as neuroscience research involves animal models that may invoke greater public concern. A few participants suggested that government policy makers should hear directly from scientists about the ongoing need for animals in research. It was noted that the patient advocacy community does not generally publicize their support of research involving animals for fear of losing donor support. As patients are the ultimate recipients of the benefits of basic and clinical science, a few participants suggested that the scientific community could engage the patient advocacy community to encourage more open discussion.

SESSION VI: CORE PRINCIPLES TO ACHIEVE CONSISTENT ANIMAL CARE AND USE OUTCOMES

Throughout the workshop many panelists and participants emphasized the need for animal research regulations that balance *quality science*, *animal welfare*, and *public confidence*. Science must be subject to strong evaluation of experiments and experimental design, observed session chair Richard Nakamura. Key animal welfare considerations are to minimize pain and distress and improve overall well-being. Many participants emphasized that public confidence stems from the assurance that animals will be appropriately protected and that science will not be inhibited from discovering much-needed solutions to global problems.

Nakamura suggested that animal welfare should not be considered in isolation from scientific goals or the larger needs of society. Several participants stressed the importance of scientific validation of animal welfare practices and standards to ensure that they actually do make a difference in terms of animal welfare. Some panelists and participants said it is also important to consider the overall costs in terms of resources on animal welfare measures.

Animal welfare issues are global, and participants urged more discussion among governments, regulators, and scientists to further the understanding of differences in regulation and impacts on animal welfare outcomes.

Participants discussed core principles for the regulation of the use of animals in research, asserting that alignment/harmonization of animal research principles may be achieved independent of differing policies or practices. Many participants believe that while there may be a need for recommendations or guidance on specific neuroscience procedures, the core principles by which animal studies should be conducted are the same for any discipline, including neuroscience.

Appendix A

References

Aldhous, P., A. Coghlan, and J. Copley. 1999. Animal experiments—where do you draw the line? Let the people speak. *New Scientist* 162(2187):26-31.

AMS (Academy of Medical Sciences). 2011. *Animals containing human material*. London, UK: The Academy of Medical Sciences.

Boomkamp, S. D., M. O. Riehle, J. Wood, M. F. Olson, and S. C. Barnett. 2012. The development of a rat in vitro model of spinal cord injury demonstrating the additive effects of Rho and ROCK inhibitors on neurite outgrowth and myelination. *Glia* 60(3):441-456.

Brennand, K. J., A. Simone, J. Jou, C. Gelboin-Burkhart, N. Tran, S. Sangar, Y. Li, Y. Mu, G. Chen, D. Yu, S. McCarthy, J. Sebat, and F. H. Gage. 2011. Modeling schizophrenia using human induced pluripotent stem cells. *Nature* 473(7346):221-225.

Courtine, G., M. B. Bunge, J. W. Fawcett, R. G. Grossman, J. H. Kaas, R. Lemon, I. Maier, J. Martin, R. J. Nudo, A. Ramon-Cueto, E. M. Rouiller, L. Schnell, T. Wannier, M. E. Schwab, and V. R. Edgerton. 2007. Can experiments in nonhuman primates expedite the translation of treatments for spinal cord injury in humans? *Nat Med* 13(5):561-566.

Crossley, N. A., E. Sena, J. Goehler, J. Horn, B. van der Worp, P. M. Bath, M. Macleod, and U. Dirnagl. 2008. Empirical evidence of bias in the design of experimental stroke studies: A meta-epidemiologic approach. *Stroke* 39(3):929-934.

European Communities and Office for Official Publications. 1986. Council directive on the approximation of laws, regulations and administrative provisions of member states regarding the protection of animals used for experimental and other scientific purposes. *Official Journal of the European Communities* No. L 358 (18. 12. 86).

European Union. 2010. Directive 2010/63/EU of the European parliament and of the council of 22 September 2010 on the protection of animals used for scientific purposes. *Official Journal of the European Union* No. L 276/33 (20. 10. 2010).

Freund, P., E. Schmidlin, T. Wannier, J. Bloch, A. Mir, M. E. Schwab, and E. M. Rouiller. 2006. Nogo-A-specific antibody treatment enhances sprouting and functional recovery after cervical lesion in adult primates. *Nat Med* 12(7):790-792.

Ipsos MORI (Market and Opinion Research International). 2000. *Animals in medicine and science*. London, UK: Ipsos MORI.

Ipsos MORI. 2008. *Doctors still top the poll as most trusted profession.* London, UK: Ipsos MORI.

Ipsos MORI. 2010. *Views on animal experimentation.* London, UK: Ipsos MORI.

Kilkenny, C., N. Parsons, E. Kadyszewski, M. F. Festing, I. C. Cuthill, D. Fry, J. Hutton, and D. G. Altman. 2009. Survey of the quality of experimental design, statistical analysis and reporting of research using animals. *PLoS One* 4(11):e7824.

Lemon, R. N. 2008. Descending pathways in motor control. *Annu Rev Neurosci* 31:195-218.

MRC (Medical Research Council). 2006. *The use of non-human primates in research—The Weatherall Report.* London, UK: MRC.

MRC. 2011. *The review of research using non-human primates: Report of a panel chaired by Professor Sir Patrick Bateson FRS.* London, UK: MRC.

NRC (National Research Council). 2003. *Guidelines for the care and use of mammals in neuroscience and behavioral research.* Washington, DC: The National Academies Press.

NRC. 2006. *Reaping the benefits of genomic and proteomic research: Intellectual property rights, innovation, and public health.* Washington, DC: The National Academies Press.

NRC. 2010. *Guide for the care and use of laboratory animals: 8th ed.* Washington, DC: The National Academies Press.

OIE (World Organisation for Animal Health). 2011. *Terrestrial Animal Health Code.* Paris, France: World Organisation for Animal Health.

Perel, P., I. Roberts, E. Sena, P. Wheble, C. Briscoe, P. Sandercock, M. Macleod, L. E. Mignini, P. Jayaram, and K. S. Khan. 2007. Comparison of treatment effects between animal experiments and clinical trials: Systematic review. *BMJ* 334(7586):197.

Russell, W., and R. Burch. 1959. *The principles of humane experimental technique.* London, UK: Methuen.

SCHER (Scientific Committee on Health and Environmental Risks). 2009. *The need for non-human primates in biomedical research, production and testing of products and devices.* Brussels, Belgium: European Commission.

Schnell, L., and M. E. Schwab. 1990. Axonal regeneration in the rat spinal cord produced by an antibody against myelin-associated neurite growth inhibitors. *Nature* 18(343):269-272.

Sloviter, R. S. 2005. The neurobiology of temporal lobe epilepsy: Too much information, not enough knowledge. *C R Biol* 328(2):143-153.

Sorensen, A., K. Moffat, C. Thomson, and S. C. Barnett. 2008. Astrocytes, but not olfactory ensheathing cells or Schwann cells, promote myelination of CNS axons in vitro. *Glia* 56(7):750-763.

Yang, S. H., P. H. Cheng, H. Banta, K. Piotrowska-Nitsche, J. J. Yang, E. C. Cheng, B. Snyder, K. Larkin, J. Liu, J. Orkin, Z. H. Fang, Y. Smith, J. Bachevalier, S. M. Zola, S. H. Li, X. J. Li, and A. W. Chan. 2008. Towards a transgenic model of Huntington's disease in a non-human primate. *Nature* 453(7197):921-924.

Zörner, B., and M. E. Schwab. 2010. Anti-Nogo on the go: From animal models to a clinical trial. *Ann NY Acad Sci* 1198(Suppl 1):E22-E34.

Appendix B

Workshop Agenda

**U.S. AND EUROPEAN ANIMAL RESEARCH REGULATIONS:
IMPACT ON NEUROSCIENCE RESEARCH**

July 26-27, 2011
Kavli Royal Society International Centre, Chicheley Hall
Buckinghamshire, United Kingdom

Background: Numerous regulations, laws, directives, and policies are in place to ensure the ethical use of animals in medical and life sciences research. These regulations are intended to ensure that humane care and use is provided to animals in research and that practical steps are taken to use the smallest number of animals to give significant results while ensuring that each individual animal experiences minimum pain or distress. The goal of the workshop is to bring together researchers, legal scholars, administrators, and other key stakeholders to discuss current trends and differences in animal regulations. Particular attention will be paid to identifying potential implications of new regulations on neuroscience research. The workshop will also provide an opportunity for international dialog about engaging public opinion regarding animal use in research and the development of core principles and outcomes for animal care and use.

Meeting Objectives:
With particular reference to neuroscience research, to:

- Identify and discuss international differences in animal research regulations:
  o   Discuss current and emerging issues.
- Discuss legal trends and activity in the courts that may impact research.
- Examine the implications of regulations on the neuroscience research enterprise.
- Discuss current communication strategies regarding animal research.
- Explore the feasibility of developing a set of global core principles and outcomes for animal care and use.

DAY ONE

8:00 a.m.   Breakfast

8:30 a.m.   Welcome, Introductions, and Objectives
              COLIN BLAKEMORE, *Co-Chair*
              ARTHUR SUSSMAN, *Co-Chair*

8:45 a.m.   Animal Research in the Neurosciences
              COLIN BLAKEMORE

SESSION I: CURRENT REGULATIONS AND EMERGING ISSUES

Session Objective: Highlight current animal research regulations, policies, and guidance. Review differences in approaches to regulations and practices exemplified by the United States and European Union and new regulations currently being proposed in emerging regions (e.g., Asia and South America). Include a review of current and emerging issues in animal research regulations.

9:15 a.m.   Overview and Session Objectives
              JUDY MacARTHUR CLARK, *Session Chair*

9:25 a.m.   Europe
              KARIN BLUMER
              Scientific Affairs
              Novartis, Switzerland

9:50 a.m.   United States
               TAYLOR BENNETT
               Senior Scientific Advisor
               National Association for Biomedical Research

10:15 a.m.   Asia
               JIANFEI WANG
               Director, Laboratory Animal Science
               GlaxoSmithKline, R&D China

10:40 a.m.   South America
               EKATERINA RIVERA
               Professor, Biological Sciences Institute
               University of Goias

11:05 a.m.   Panel Discussion with Speakers and Participants:
               • What is the basis for regulatory differences among
                 countries?
               • What are emerging key issues surrounding animal re-
                 search regulations?

12:00 p.m.   LUNCH

SESSION II: IMPACT OF LEGAL TRENDS ON ANIMAL RESEARCH

Session Objective: Discuss changes to laws regarding animal rights on regulations and research. Explore emerging laws and legal strategies that have
the potential to directly influence the use of animals in medical research.

1:15 p.m.   Overview and Session Objectives
               ARTHUR SUSSMAN, *Session Chair*

1:25 p.m.   Animal Rights Laws
               • Examine the interplay between the legal system and animal
                 use regulatory system.
               • What are the potential implications of changes in regula-
                 tions to the legal rights of animals? Research?
               • What are current trends, United States versus European
                 differences?

               MARGARET FOSTER RILEY
               Professor
               University of Virginia, School of Law

1:45 p.m.   Freedom of Information and Openness
- How are these laws being used by the animal rights movement?
- Are there limitations to the type of information that can be obtained?
- Can greater transparency lead to less effort needed in response to Freedom of Information Act demands?

> MARGARET SNYDER
> Freedom of Information Act Coordinator
> Office of Extramural Research
> National Institutes of Health

2:05 p.m.   State Sunshine Laws
- How are these laws being used by the animal rights movement?
- Are there limitations to the type of information that can be obtained?
- Can greater transparency reduce requests for information?

> RICHARD CUPP
> John W. Wade Professor of Law
> Pepperdine Law School

2:15 p.m.   Discussion with Speakers and Participants

2:40 p.m.   BREAK

SESSION III: THE IMPACT OF REGULATIONS ON ANIMAL-BASED NEUROSCIENCE RESEARCH

Session Objective: Discuss the impact of current regulations, policies, guidance, and economic considerations on the conduct of animal-based neuroscience research. Consider the role that animals have played in neuroscience research: the benefits achieved, but also the costs. Include examination of the administrative load and economic cost associated with animal research regulations and response of researchers and funders to cost implications.

3:10 p.m.   Overview and Session Objectives
ROBERTO CAMINITI, *Session Chair*

3:20 p.m.   Panelists:

Use of Rodent Models in Neuroscience
BILL YATES
Professor
University of Pittsburgh

When Should Non-Human Primates Be Used as Animal Models?
ROGER LEMON
Sobell Chair of Neurophysiology
University College London Institute of Neurology

The Ethical and Practical Dilemmas of Research on Non-Human Primates
STUART ZOLA
Director
Yerkes National Primate Research Center

Administrative and Economic Costs
CHARLES J. HECKMAN
Professor
Northwestern University Feinberg School of Medicine

4:40 p.m.   Discussion with Speakers and Participants:
• How has the implementation of current and new regulations impacted the speed and quality of research, positively and negatively?
• Has the pressure for reduction of numbers, use of "lower" species, reduction of cost, and replacement of animals distorted the balance of neuroscience research in ways that impede the rate of discovery?
• How can we assess costs to animals, especially cumulative severity in long-term animal studies, including NHPs?
• How can administrators and scientists work together to balance the economic costs of animal research regulations while maintaining public confidence?

5:30 p.m.   ADJOURN AND DINNER IN THE HALL'S DINING ROOM

DAY TWO

8:00 a.m.   Breakfast

SESSION IV: IMPACT OF 3Rs ON THE
NEUROSCIENCE RESEARCH ENTERPRISE

Session Objective: Examine experiences of applying the 3Rs (replacement, refinement, and reduction) in neuroscience research, including consideration of opportunities for enhanced scientific outcomes as well as welfare benefits and potential limitations. Examine the influence that non-researchers and others have on neuroscience researchers working with animals. Consider the role of systematic reviews, or the review and synthesis of all relevant studies by the application of scientific strategies.

8:30 a.m.   Overview and Session Objectives
             COLIN BLAKEMORE, *Session Chair*

8:40 a.m.   Panelists:

             Replacement Strategies in Neuroscience Research: Focus
             on Spinal Cord Injury
                SUE BARNETT
                Professor of Cellular Neuroscience
                University of Glasgow

             Refinement and Reduction Strategies: Improving Models
             of Disease and Using Translational Approaches in Epilepsy
             and Parkinson's Disease
                GAVIN WOODHALL
                Reader in Neuropharmacology
                Aston University

             The Role of Systematic Reviews
                ANNE MURPHY
                Associate Professor
                University of California, San Diego

             Future Considerations and Impact of 3Rs
                JACKIE HUNTER
                OI Pharma Partners

10:20 a.m. Discussion with Speakers and Participants:
- How can the 3Rs best be used effectively to deliver advancements in neuroscience?
- For what areas of neuroscience research is replacement a realistic long-term goal? How can this objective be most effectively pursued?
- Are current regulations causing neuroscientists to move away from animal work or to use less strictly regulated models?
- Are new regulations impeding the progress of neuroscience, or leading to neuroscience advancements?
- Is collaboration between sectors (industry/academia) effective and what is the impact of greater globalization of research?
- Critical analysis of systematic reviews—do they play a role? If so, should there be a new approach to experimental design to facilitate such reviews?

11:00 a.m. BREAK

SESSION V: ENGAGING AND INFORMING THE PUBLIC

Session Objective: Provide an opportunity for international dialog around communication strategies regarding animal use in research. Examine successes and failures in the engagement of the public, politicians, and the media in productive discussions of the use of animals in research. Identify opportunities to educate non-researchers in the animal use regulatory system.

11:15 a.m. Overview and Session Objectives
FRANKIE TRULL, *Session Chair*

11:25 a.m. Panelists:

Neuroscientist
RANDALL NELSON
Professor
The University of Tennessee Health Science Center

Science Writer
MARK HENDERSON
Science Correspondent
The Times

Patient Group Administrator
TIM COETZEE
Chief Research Officer
National Multiple Sclerosis Society

12:25 p.m.   Discussion with Speakers and Participants:
  • What is the responsibility of individual scientists, patient groups, and organizations to engage the public in dialogue about animal research?
  • Are there teachable examples of successful engagement and dialogue by animal researchers with the public?

1:00 p.m.   LUNCH

SESSION VI: CORE PRINCIPLES FOR ANIMAL RESEARCH REGULATION

Session Objective: Provide an opportunity for international dialog around the development of core principles and outcomes for regulating animal research. Identify areas of research where such adoption would be beneficial. Discuss next steps in development of core principles and outcomes, including analysis of the role of the 3Rs (replacement, reduction, and refinement). Identify key stakeholders important for the success of this endeavor.

2:00 p.m.   Overview and Session Objectives
RICHARD NAKAMURA, *Session Chair*

2:10 p.m.   Panelists:

European Government Regulator
JUDY MacARTHUR CLARK
Chief Inspector
UK Home Office

U.S. Government Regulator
PATRICIA BROWN
Director
Office of Laboratory Animal Welfare

Industry Representative
MARGARET LANDI
Vice President, Global Laboratory Animal Science;
Chief of Animal Welfare and Veterinary Medicine
GSK Pharmaceuticals

ILAR Council Member
TIMO NEVALAINEN
Professor
University of Eastern Finland

3:30 p.m.   Discussion with Speakers and Participants:
 • Are there core principles and outcomes specific to regulations for animal use in neuroscience research?

4:15 p.m.   MEETING WRAP-UP WITH SESSION CHAIRS

Panelists:
Session I: JUDY MacARTHUR CLARK
Session II: ARTHUR SUSSMAN
Session III: ROBERTO CAMINITI
Session IV: COLIN BLAKEMORE
Session V: FRANKIE TRULL
Session VI: RICHARD NAKAMURA

5:00 p.m.   FINAL REMARKS
COLIN BLAKEMORE, *Co-Chair*
ARTHUR SUSSMAN, *Co-Chair*

5:30 p.m.   ADJOURN

Appendix C

Registered Attendees

Neeraj Agarwal
National Institutes of Health

Caroline Bergmann
University of Oxford

Lynsey Bilsland
Wellcome Trust

Jennifer Bizley
University College London

Laura Boothman
Academy of Medical Sciences

Victoria Cambridge
The Royal Society

Kate Chandler
UK Home office

Linda Chezem
Purdue University and Indiana
University

Beverley Clark
University College London

Anne Deschamps
Federation of American Societies
for Experimental Biology

Paul Flecknell
Newcastle University

Richard Fosse
GlaxoSmithKline

Lee Glassbrook
Biotechnology and Biological
Sciences Research Council

Jon Hatcher
MedImmune

Robert Hubrecht
Universities Federation for Animal
Welfare

Maggy Jennings
The Royal Society for the Prevention
   of Cruelty to Animals

Sharon Juliano
Uniformed Services University of
   the Health Sciences

Martin Lawton
University College London

Nancy Lee
Wellcome Trust

Kirk Leech
Institute of Ideas

Daniel Marsman
American Veterinary Medical
   Association

Paul Matthews
GlaxoSmithKline

Emily McIvor
The Humane Society International

Sarah Mee
The Royal Society

Anna Mitchell
Oxford University

Gianni Dal Negro
GlaxoSmithKline

John O'Keefe
University College London

Chris Powell
GlaxoSmithKline

Frances Rawle
Medical Research Council

Janet Rodgers
University of Oxford

Claire Russell
Royal Veterinary College

Stephen Ryder
UK Home Office

Richard Saunders
National Institute of Mental
   Health

David Shurtleff
National Institute on Drug Abuse

Christopher Wathes
Farm Animal Welfare Committee

Nicola Watts
AstraZeneca UK Veterinary Group

Martin Whiting
Royal Veterinary College

David Whittaker
Oxford University

Robin Williams
Royal Holloway University of
   London

Sarah Wolfensohn
UK Animal Procedures Committee